GOVERNMENT PROCEDURES AND OPERATIONS

FEDERAL DISPOSAL RESPONSIBILITIES

SELECTED ISSUES, ASSETS, AND ANALYSES

GOVERNMENT PROCEDURES AND OPERATIONS

Additional books in this series can be found on Nova's website under the Series tab.

Additional e-books in this series can be found on Nova's website under the e-book tab.

GOVERNMENT PROCEDURES AND OPERATIONS

FEDERAL DISPOSAL RESPONSIBILITIES

SELECTED ISSUES, ASSETS, AND ANALYSES

TARA C. WALLER
EDITOR

nova publishers
New York

For permission to use material from this book please contact us:
Telephone 631-231-7269; Fax 631-231-8175
Web Site: http://www.novapublishers.com

Library of Congress Cataloging-in-Publication Data

ISBN: 978-1-63321-217-6

Published by Nova Science Publishers, Inc. † New York

CONTENTS

PREFACE

Real property disposal is the process by which federal agencies identify and then transfer, donate, or sell real property they no longer need. Disposition is an important asset management function because the costs of maintaining unneeded properties can be substantial, consuming financial resources that might be applied to long-standing real property needs, such as repairing existing facilities, or other pressing policy issues, such as reducing the national debt. Also, timely and proper disposal of obsolete ships in the National Defense Reserve Fleet, is critical to protecting the environment. Because these ships often contain hazardous materials, members of Congress and others have raised issues about the environmental concerns. This is also true for the management of civilian radioactive waste, and defense surplus equipment disposal.

Chapter 1 – Real property disposal is the process by which federal agencies identify and then transfer, donate, or sell real property they no longer need. Disposition is an important asset management function because the costs of maintaining unneeded properties can be substantial, consuming financial resources that might be applied to long-standing real property needs, such as repairing existing facilities, or other pressing policy issues, such as reducing the national debt.

Despite the expense, federal agencies hold thousands of unneeded and underutilized properties. Agencies have argued that they are unable to dispose of these properties for several reasons. First, there are statutorily prescribed steps in the disposal process that can take months to complete. Second, agencies are often required to complete major repairs or environmental remediation before properties are ready for disposal—steps for which agencies lack funding. Third, key stakeholders in the disposal process—including local governments, non-profit organizations, and businesses—are often at odds over how to dispose of properties. In addition, Congress may be limited in its capacity to conduct oversight of the disposal process because it currently lacks access to reliable, comprehensive real property data.

Three bills have been introduced in the 113th Congress that propose significant changes to the existing real property disposal system. The Federal Real PropertyAsset Management ReformAct of 2013 (S. 1398), would establish an expedited disposal program under which 200 properties would be exempt from time-consuming, statutory disposal requirements. In addition, S. 1398 would expand the role of an interagency workgroup, the Federal Real Property Council, to set disposal goals for agencies and monitor their progress in meeting those goals. The bill would also increase oversight of agency disposal activities by requiring the Administrator of the General Services Administration (GSA) to establish a real property database available to the public at no cost.

The Excess Federal Building and Property Disposal Act of 2013 (H.R. 328) would establish an expedited disposal program under which the 15 unneeded federal properties with the highest fair market value would bypass statutory disposal requirements and be offered for sale immediately. H.R. 328 would also require the GSA Administrator to establish a real property database available to the public at no cost and provide a report to Congress on the progress each landholding agency has made in reducing its unneeded property.

The Civilian Property Realignment Act (H.R. 695) would centralize the disposal process by establishing a Civilian Property Realignment Commission, which would work with agencies to develop a list of disposal recommendations to the President. If the President approved the recommendations, then they would be sent to Congress. If Congress passed a joint resolution of approval then agencies would be required to implement the recommendations; if a joint resolution of approval was not passed then the realignment process would end for the fiscal year.

Chapter 2 – Timely and proper disposal of obsolete ships in the National Defense Reserve Fleet—older ships designated for use in national emergencies—is critical to protecting the environment. Because these ships often contain hazardous materials, members of Congress and others have raised issues about the environmental concerns. As part of the Department of Transportation, MARAD's Ship Disposal Program serves as the federal government's agent for competing and awarding contracts for recycling the ships' materials. Congress has required MARAD's ship disposal program to award ship recycling contracts to qualified ship recycling facilities on the basis of best value. The Coast Guard and Maritime Transportation Act of 2012 mandated that GAO review MARAD's source selection procedures and practices used to award ship recycling contracts.

In this report, GAO assessed MARAD's (1) source selection process; (2) communication strategy with ship recycling contractors; and (3) long-term ship disposal strategy. To complete this work, GAO reviewed and analyzed documentation on MARAD's qualification process, source selection procedures, and strategies; and interviewed MARAD and all of its qualified ship recycling contractors.

Chapter 3 – Management of civilian radioactive waste has posed difficult issues for Congress since the beginning of the nuclear power industry in the 1950s. Federal policy is based on the premise that nuclear waste can be disposed of safely, but proposed storage and disposal facilities have frequently been challenged on safety, health, and environmental grounds. Although civilian radioactive waste encompasses a wide range of materials, most of the current debate focuses on highly radioactive spent fuel from nuclear power plants. The United States currently has no disposal facility for spent nuclear fuel.

The Nuclear Waste Policy Act of 1982 (NWPA) calls for disposal of spent nuclear fuel in a deep geologic repository. NWPA established the Office of Civilian Radioactive Waste Management (OCRWM) in the Department of Energy (DOE) to develop such a repository, which would be licensed by the Nuclear Regulatory Commission (NRC). Amendments to NWPA in 1987 restricted DOE's repository site studies to Yucca Mountain in Nevada. DOE submitted a license application for the proposed Yucca Mountain repository to NRC on June 3, 2008. The state of Nevada strongly opposes the Yucca Mountain project, citing excessive water infiltration, earthquakes, volcanoes, human intrusion, and other technical flaws.

The Obama Administration "has determined that developing the Yucca Mountain repository is not a workable option and the Nation needs a different solution for nuclear waste disposal," according to the DOE FY2011 budget justification. As a result, no funding for Yucca Mountain, OCRWM, or NRC licensing was requested or provided for FY2011 or subsequent years. DOE filed a motion with NRC to withdraw the Yucca Mountain license application on March 3, 2010. An NRC licensing board denied DOE's withdrawal motion on June 29, 2010, a decision sustained by the NRC commissioners on a tie vote September 9, 2011. Despite that decision, NRC halted further consideration of the license application because of "budgetary limitations," but a federal appeals court on August 13, 2013, ordered NRC to continue the licensing process with previously appropriated funds.

After halting the Yucca Mountain project, the Administration established the Blue Ribbon Commission on America's Nuclear Future to develop an alternative nuclear waste policy. The commission issued its final report on January 26, 2012, recommending that a new, "single-purpose organization" be given the authority and resources to promptly begin developing one or more nuclear waste repositories and consolidated storage facilities. The commission recommended a "consent based" process for siting nuclear waste storage and disposal facilities and that long-term research, development, and demonstration be conducted on technologies that could provide waste disposal benefits.

After OCRWM was dismantled, responsibility for implementing the Administration's nuclear waste policy was given to DOE's Office of Nuclear Energy (NE). In January 2013, NE issued a nuclear waste strategy based on the Blue Ribbon Commission recommendations. The strategy calls for a pilot interim storage facility for spent fuel from closed nuclear reactors to open by 2021 and a larger storage facility, possibly at the same site, to open by 2025. A site for a permanent underground waste repository would be selected by 2026, and the repository would open by 2048. DOE requested $60 million for FY2014 to carry out the new waste strategy, an amount approved by the Senate Appropriations Committee. However, the House denied funding for the new strategy and provided $25 million for Yucca Mountain instead.

Chapter 4 – The Department of Defense (DOD) through a Defense Logistics Agency (DLA) component called DLA Disposition Services [formerly the Defense Utilization and Marketing Service (DRMS)] has a policy for disposing of government equipment and supplies considered surplus or deemed unnecessary, or excess to the agency's currently designated mission. DLA Disposition Services is responsible for property reuse (including resale), precious metal recovery, recycling, hazardous property disposal, and the demilitarization of military equipment. The effort to dispose of surplus military equipment dates back to the end of World War II when the federal government sought to reduce a massive inventory of surplus military equipment by making such equipment available to civilians. (The disposal of surplus real property, including land, buildings, commercial facilities, and equipment situated thereon, is assigned to the General Services Administration, Office of Property Disposal.)

Chapter 5 – The Defense Base Closure and Realignment Act of 1990 (P.L. 100-526) and the Federal Property and Administrative Services Act of 1949

(P.L. 81-152) provide the basic framework for the transfer and disposal of military installations closed during the base realignment and closure (BRAC) process. In general, property at BRAC installations is first subjected to screening for use by the Department of Defense and by other federal agencies. If no federal use for the property can be found or if an application for transfer is rejected, the property is deemed "surplus" to the needs of the federal government and made available for disposal through other mechanisms.

At this point, BRAC property is subjected to two simultaneous evaluation processes: the redevelopment planning process performed by a local redevelopment authority composed of various interested representatives of the community affected by the BRAC action; and a Department of Defense analysis prepared under the aegis of the National Environmental Policy Act and, eventually, informed by the local redevelopment plan.

As a part of this process, screening of the property must be performed to determine if a homeless assistance use would be appropriate. There are also a variety of "public benefit transfers," under which the property may be conveyed for various specified public purposes at reduced cost. It is also possible to dispose of BRAC property through the use of a public auction or negotiated sale, for which fair market value or a proxy for fair market value must generally be obtained. Finally the law governing the BRAC process authorizes economic development conveyances, through which a local redevelopment authority may obtain the property for specified purposes, sometimes for no consideration.

The BRAC property transfer process has been altered, both legislatively and administratively, throughout the numerous authorized closure rounds. Most recently, the National Defense Authorization Act for Fiscal Year 2010 (P.L. 111-84) amended the law with respect to economic development conveyances at no cost to local redevelopment authorities. This report provides an overview of the various authorities available under the current law and describes the planning process for the redevelopment of BRAC properties.

In: Federal Disposal Responsibilities
Editor: Tara C. Waller
ISBN: 978-1-63321-217-6

Chapter 1

DISPOSAL OF UNNEEDED FEDERAL BUILDINGS: LEGISLATIVE PROPOSALS IN THE 113TH CONGRESS*

Garrett Hatch

SUMMARY

Real property disposal is the process by which federal agencies identify and then transfer, donate, or sell real property they no longer need. Disposition is an important asset management function because the costs of maintaining unneeded properties can be substantial, consuming financial resources that might be applied to long-standing real property needs, such as repairing existing facilities, or other pressing policy issues, such as reducing the national debt.

Despite the expense, federal agencies hold thousands of unneeded and underutilized properties. Agencies have argued that they are unable to dispose of these properties for several reasons. First, there are statutorily prescribed steps in the disposal process that can take months to complete. Second, agencies are often required to complete major repairs or environmental remediation before properties are ready for disposal—steps for which agencies

* This is an edited, reformatted and augmented version of a Congressional Research Service publication R43247, prepared for Members and Committees of Congress dated September 27, 2013.

lack funding. Third, key stakeholders in the disposal process—including local governments, non-profit organizations, and businesses—are often at odds over how to dispose of properties. In addition, Congress may be limited in its capacity to conduct oversight of the disposal process because it currently lacks access to reliable, comprehensive real property data.

Three bills have been introduced in the 113th Congress that propose significant changes to the existing real property disposal system. The Federal Real PropertyAsset Management ReformAct of 2013 (S. 1398), would establish an expedited disposal program under which 200 properties would be exempt from time-consuming, statutory disposal requirements. In addition, S. 1398 would expand the role of an interagency workgroup, the Federal Real Property Council, to set disposal goals for agencies and monitor their progress in meeting those goals. The bill would also increase oversight of agency disposal activities by requiring the Administrator of the General Services Administration (GSA) to establish a real property database available to the public at no cost.

The Excess Federal Building and Property Disposal Act of 2013 (H.R. 328) would establish an expedited disposal program under which the 15 unneeded federal properties with the highest fair market value would bypass statutory disposal requirements and be offered for sale immediately. H.R. 328 would also require the GSA Administrator to establish a real property database available to the public at no cost and provide a report to Congress on the progress each landholding agency has made in reducing its unneeded property.

The Civilian Property Realignment Act (H.R. 695) would centralize the disposal process by establishing a Civilian Property Realignment Commission, which would work with agencies to develop a list of disposal recommendations to the President. If the President approved the recommendations, then they would be sent to Congress. If Congress passed a joint resolution of approval then agencies would be required to implement the recommendations; if a joint resolution of approval was not passed then the realignment process would end for the fiscal year.

BACKGROUND

Federal executive branch agencies hold an extensive real property portfolio that includes approximately 399,000 buildings. These assets have been acquired over a period of decades to help agencies fulfill their diverse missions. Agencies hold buildings with a range of uses, including offices, health clinics, warehouses, and laboratories. As agencies' missions change over time, so, too, do their real property needs, thereby rendering some assets less useful or unneeded altogether. Healthcare provided by the Department of VeteransAffairs (VA) has shifted in recent decades from predominately hospital-based inpatient care to a greater reliance on clinics and outpatient care, with a resulting change in space needs. Similarly, the Department of Defense (DOD) reduced its force by 36% after the Cold War ended, and has engaged in several rounds of base realignments and installation closures.

Agencies are required to dispose of real property that they no longer need, but many continue to hold onto unneeded building space. In FY2010, the government held 77,700 buildings it identified as either not utilized or underutilized and spent $1.67 billion operating and maintaining them.[1] Federal agencies have indicated that their disposal efforts are often hampered by legal and budgetary disincentives, and competing stakeholder interests.[2]

This report begins with an explanation of the real property disposal process and then discusses some of the factors that have made disposition relatively inefficient and costly. It then examines key provisions of three real property reform bills introduced in the 113th Congress: the Federal Real PropertyAsset Management ReformAct of 2013 (S. 1398); the Excess Federal Building and Property Disposal Act of 2013 (H.R. 328); and the Civilian Property Realignment Act of 2013 (H.R. 665). This report concludes with a discussion of policy options for enhancing the disposal process, including the potential use of Public-Private Partnerships to generate revenue from underutilized properties.

Obstacles to Timely and Efficient Disposition

As noted, the government maintains a large inventory of unneeded or underutilized properties. These properties not only incur costs to the government to operate and maintain, but could, in some instances, be utilized by nonfederal entities—state and local governments, nonprofits, private sector businesses—to accomplish a range of public purposes, such as providing services to the homeless, or facilitating economic development. The Government Accountability Office (GAO) reports have consistently noted that efforts to dispose of unneeded and underutilized properties are hindered by statutory disposal requirements, the cost of preparing properties for disposal, conflicts with stakeholders, and a lack of accurate data. Each of these issues is discussed here.

Identifying Unneeded Space

Agencies are required to continuously survey property under their control to identify any property that it no longer needs to carry out its mission—excess property—and to "promptly" report that property as excess to the General Services Administration (GSA).[3] Agencies are then required to follow the regulations prescribed by GSA when disposing of unneeded property or to follow independent or delegated statutory authority.[4] GSA's regulations, in turn, implement statutory disposal requirements, discussed below.[5]

Statutory Disposal Requirements

The steps in the real property disposal process are set by statute. Agencies must first offer to transfer properties they do not need (excess properties) to other federal agencies, who generally pay market value for excess properties they wish to acquire.[6] Unneeded properties that are not acquired by federal agencies (surplus properties) must then be offered to state and local governments, and qualified nonprofits, for use in accomplishing public purposes specified in statute, such as use as public parks or for providing services to the homeless.[7] Agencies may convey surplus properties to state and local governments, and qualified nonprofits, for public benefit at less than fair market value—even at no cost.[8] Surplus properties not conveyed for public benefit are then available for sale at fair market value or are demolished if the property could not be sold due to the condition or location of the property.[9]

Agencies have consistently argued that these statutory requirements slow down the disposition process, compelling agencies to incur operating costs for months—sometimes years—while the properties are being screened.[10] Real property officials at the VA have said the McKinney-Vento Act (P.L. 100-77)—which mandates that all surplus property be screened for homeless use—can add as much as two years to the disposal process.[11] Because public benefit conveyance requirements are set in law, agencies do not have the authority to skip screening, even for surplus properties that could not be conveyed anyway. The Department of Energy (DOE), for example, told auditors that they had properties that they felt could be disposed of only by demolition, due to their condition or location, but that still had to go through the screening process, thereby forcing DOE to pay maintenance costs that could have been avoided.[12]

Statutes pertaining to environmental remediation or historic preservation also add time to the process. It may take agencies years of study to assess the potential environmental consequences of a proposed disposal and to develop and implement an abatement plan, as required by law.[13] Similarly, the National Historic Preservation Act[14] requires agencies to plan their disposal actions so as to minimize the harm they cause to historic properties, which may include additional procedures, such as consulting with historic preservation groups at the state, local, and federal level.[15]

Disposal Costs

Unneeded buildings are often among the older properties in an agency's portfolio. As a consequence, agencies sometimes find that they are required to complete expensive repairs and renovations before the properties are ready for disposal. Agencies may need to invest in repairs that will enable a building to meet health and safety standards, for example, or restore historic sites in accordance with federal standards. It has been estimated, for example, that VA would need to spend about $3 billion to repair the buildings in its portfolio that it rated in "poor" or "critical" condition—56% of which were vacant or underutilized, and therefore might be candidates for disposal.[16] Agencies that wish to demolish vacant buildings face deconstruction and cleanup costs that, at times, exceed the cost of maintaining the property—at least in the short run—which may encourage real property managers to retain a property rather than dispose of it.[17] Federal agencies frequently cite the cost of complying with environmental regulations as a major disincentive to disposal.[18] Generally speaking, agencies are required to assess and pay for any environmental cleanup that may be needed before disposing of a property. Identifying and addressing environmental hazards, such as lead paint, asbestos, medical waste, and soil contamination, prior to disposition can result in "significant" up-front costs for agencies.[19]

Stakeholder Conflict

Some agencies have found their disposal efforts complicated by the involvement of stakeholders with competing agendas. The Department of the Interior (DOI) has said that it can be stymied by the competing concerns of local and state governments, historic preservation offices, and political factors, when attempting to dispose of some of its unneeded real property.[20] Similarly, VA has found that communities sometimes oppose disposals that would result in new development, and veterans groups have opposed disposing of building space if that space would be used for purposes unrelated to the needs of veterans.[21] The Department of State (DOS) has had difficulty in disposing of surplus real property overseas due to disputes with host governments that restrict property sales.[22] These conflicts can result in delay, or even cancellation of proposed disposals, which, in turn, prevent agencies from reducing their inventories of unneeded properties.[23]

Real Property Management and Oversight

In addition to the obstacles mentioned above, data about agency real property portfolios—which might be useful for congressional oversight—appear to be inaccurate, and government-wide data are accessible only to the agency that manages the database, the GSA. Moreover, agencies regularly enter into leases rather than seek funding for new construction when acquiring space, even when the leased space is more expensive over time.

Availability and Quality of Real Property Data

The Federal Real Property Profile (FRPP) is the government's most comprehensive source of information about real property under the control of executive branch agencies. GSA manages the FRPP and collects real property data from 24 of the largest landholding agencies each year. Other agencies are encouraged, but not required, to report data to GSA.[24] The data elements that participating agencies collect and report are determined by the Federal Real Property Council (FRPC), an interagency taskforce that is funded and chaired by the Office of Management and Budget (OMB). The other members of the FRPC are agency senior real property officers (SRPOs) and GSA.

The FRPP contains data that could enhance congressional oversight of federal real property activities, such as the number of excess and surplus properties held by major landholding agencies, the annual costs of maintaining those properties, and agency disposition actions. GSA, however, does not permit direct access to the FRPP by Congress on the grounds that the data are proprietary. GSA does respond to requests for real property data from congressional offices, but GSA staff query the database and provide the results to the requestor.

Some FRPP data are made public through an annual summary report posted on GSA's website, but the summary reports are of limited use for congressional oversight for several reasons.[25] Most of the data are highly aggregated (e.g., the number of assets disposed of, government-wide, through public benefit conveyance), and very limited information is provided on an agency-by-agency basis. It is not possible, therefore, for Congress to monitor the performance of individual agencies through the summary reports. Basic questions, such as how many excess and surplus properties each agency holds or has disposed of in a given fiscal year, cannot be answered. Nor is it possible to compare the performance of agencies, which limits the ability of Congress to study the policies and practices at the most successful agencies and hold poorly performing agencies accountable.

The quality of the FRPP data has also been questioned. GAO audits have found, for example, that some real property data were incomplete or were not comparable across agencies, which limited the usefulness of those data for analysis.[26] The most recent GAO report, from June 2012, declared that the FRPP had not been populated through sound data collection practices and key data elements—such as a building's utilization, condition, annual operating costs, mission dependency, and value—are not consistently and accurately captured in the database.[27] The GAO report concluded that FRPP users "cannot be sure that the data are sufficiently reliable to support sound management and decision making about excess and underutilized property."[28]

In addition, annual summary reports based on FRPP data may miscategorize important information on disposal methods. As discussed previously, agencies are statutorily required to dispose of properties through transfer, conveyance, sale, or demolition. Recently published FRPC summary reports, however, identify "other" as the largest or second largest category of property disposal, accounting for 46% of the total number of real property assets disposed by agencies in FY2007, nearly 73% of those disposed in FY2008, 41% in FY2009, and 33% in FY2010.[29] Typically, the "other" data category is reserved for a relatively small number of cases that do not clearly fit into one of the major data categories, so it is unusual to see such a large number of "other" dispositions. In fact, the FRPP defines "other" disposals as those "that cannot be classified in any of the other disposition methods." The annual reports, however, do not explain why so many disposals cannot be classified as transfer, conveyance, sale, or demolition.[30] One explanation may be that agencies are misreporting their disposal data; another may be that some disposals are a combination of methods. If so, then the data reported for all types of dispositions may be of limited use, because thousands of properties may have been miscategorized.

The annual summary reports also omit data that might enhance congressional oversight. The FRPP contains, for example, the number of excess and surplus properties held by each agency and the annual operating costs of those properties—issues about which Congress has expressed ongoing interest—but the summary report only provides the number and annual operating costs of disposed assets, thereby providing the "good news" of future costs avoided through disposition while omitting the "bad news" of the ongoing operating costs associated with excess and surplus properties the government maintained. In addition, agencies estimate a dollar amount for the repair needs of their buildings and structures as part of their FRPP reporting, but the estimate is then folded into a formula for calculating the condition of each building.[31] Given that repair needs are an obstacle to disposing of some properties, Congress may find it useful to have the repair estimates reported separately to help inform funding decisions.

Overreliance on Leasing

In a 2011 report, GAO wrote that it considers the government's "overreliance on costly leased space" to be one of the primary reasons federal real property continues to be designated as a "high risk" issue.[32] The percentage of square feet leased by GSA—which leases property for itself and on behalf of many agencies—now exceeds the percentage of square feet it owns. According to GAO, leasing space is typically more expensive than owning space over the same time period. GAO cited, for example, a long-term operating lease that cost an estimated $40.3 million more than if the agency had purchased the same building.[33] Similarly, in FY2010, the annual operating cost for a square foot of space in a building owned by the government was $5.30, but for leased space it was $15.00.[34]

GAO wrote that while the decision to lease rather than purchase space may be driven by operational requirements—such as the United States Postal Service (USPS) leasing space in areas that it believes will optimize the efficiency of mail delivery—agencies often choose to lease rather than purchase space because of budget scoring rules, even if the decision to lease is not the most cost-effective option. Under the Budget Enforcement Act of 1990, an agency must have budget authority up-front for the government's total legal commitment before acquiring space. Thus, if an agency were to construct or purchase a building, it would need up-front funding for the entire cost of the construction or acquisition, while leased space only requires the annual lease payment plus the cost of terminating the lease agreement.

In addition to the budget scoring issue, some agencies have been granted independent leasing authority, which means they do not have to work with GSA to acquire leased space. Some agencies with independent leasing authority, such as the USPS and VA, have established in-house real property expertise, while other agencies with independent authority have not. The Securities and Exchange Commission (SEC), for example, entered into a $557 million, 10-year lease for 900,000 square feet, which the SEC's inspector general (IG) called "another in a long history of missteps and misguided leasing decisions made by the SEC since it was granted independent leasing authority."[35] The IG found that "inexperienced senior management" at the SEC made poor decisions that led to acquiring three times the space needed—the original estimate provided to Congress was for 300,000 square feet—and bypassing other locations that were closer and less expensive.[36]

S. 1398: Federal Real Property Asset Management Reform Act of 2013

S. 1398 was introduced on July 30, 2013, and referred to the Homeland Security and Governmental Affairs Committee. The committee ordered the bill reported without amendment favorably on July 31, 2013. S. 1398 takes a broad approach to real property management, one that builds on existing resources and expertise, requires new performance measures and reporting, emphasizes finding opportunities for agency consolidation, colocation, and reconfiguration, and requires a thorough examination of the federal leasing process. It would expand the role of the FRPC in collecting and analyzing real property data. S. 1398 would also require the establishment of a real property database that might enhance congressional oversight.

Scope

S. 1398 applies to all federal agencies, which it defines as executive branch agencies and wholly owned government corporations. S. 1398 refers throughout to "underutilized" properties, which it defines a property that is used irregularly or intermittently by a federal agency or a property where a federal agency only needs a portion of the property for program purposes. The bill also defines "excess" properties as those that agency heads determine are not required to meet the needs of the agencies that control them. S. 1398 defines "surplus" properties as those that are not needed by any federal agency. Certain properties are exempt from the definition of "surplus" properties, including DOD properties subject to Base Realignment and Closure (BRAC) legislation, properties excluded for reasons of national security, certain Indian and native Eskimo lands, properties operated and maintained by the Tennessee Valley Authority, and properties operated and maintained by the USPS. The bill does not appear to exclude properties held by federal agencies in foreign countries.

Duties of Federal Agencies

S. 1398 would require federal agencies to develop a system of managing their real property holdings which would include

- maintaining adequate inventory controls and accountability systems;
- defining future workforce projections and their real property needs;
- identifying excess and underutilized properties that could be used for colocation with other federal agencies or consolidation with other facilities;
- reporting excess and underutilized property to GSA promptly;
- establishing goals for reducing underutilized property;
- submitting a report to the FRPC on all excess and underutilized properties, including an assessment of whether underutilized properties could be better utilized by the agency that controls it;
- adopting workplace practices, management techniques, and space configurations that decrease the need for space; and
- identifying underutilized leased space.

Section 629(9)(B) of the bill would also require each agency to provide an annual assessment of its real property inventory to the FRPC and the GSA Administrator, including an assessment of each property that must include the

- age and condition of the property;
- size of the property in square footage and acreage;
- geographical location of the property, including its address;
- extent to which the property is being utilized;
- actual annual operating costs associated with the property;
- total cost of capital expenditures associated with the property;
- sustainability metrics associated with the property;
- number of federal employees and functions housed at the property;
- extent to which the mission of the federal agency is dependent on the property; and
- the estimated amount of capital expenditures needed to maintain and operate the property over the next five years.

Some of these duties are already required by regulation, but, by enacting them into law, agencies would have clear standards, set in statute, against which their real property management practices could be evaluated. In addition, the language makes it clear that a complete asset management plan must include identifying opportunities for reconfiguration that could result in a more efficient use of space. Agencies would also be reporting new data that may help policymakers plan, coordinate, and execute real property disposals in the most cost-effective manner.

Duties of the Federal Real Property Council

S. 1398 would set in statute an interagency real property working group, the FRPC, which was initially established through Executive Order 13327 under President George W. Bush. S. 1398 would not alter the structure of the working group, which would consist of the SRPO of each major landholding agency, the Office of Management and Budget's (OMB's) Deputy Director for Management—who would also chair the FRPC, OMB's Controller, the GSA Administrator, and other employees the chairperson determines to be necessary. S. 1398 would also require the chairperson to designate a full-time executive director with a background in commercial real estate, real property management, and federal operations and management, to help carry out the duties of the FRPC.

The FRPC would also be required to establish a management plan template that includes performance measures, specific milestones, measurable savings, strategies and government-wide goals for reducing surplus property and improving utilization of properties that are underutilized. The FRPC would also be required to

- develop standard use rates consistent with non-governmental space use rates;
- develop a strategy to reduce reliance of federal agencies on leased space for long-term needs when ownership would be less costly;
- provide guidance on eliminating inefficiencies in the leasing process; and
- compile a list of real property assets that are field offices suitable for colocation.

In addition, the FRPC would be required to submit an annual report to the OMB Director that included a list of the remaining excess property, surplus property, and underutilized property of each federal agency. The report would also include a description of the progress the FRPC has made in fulfilling its requirements under S. 1398 and the progress agencies have made toward achieving their real property goals. In executing its duties, the FRPC would be required to consult with state and local governments, tribal authorities, and affected communities, as well as private sector entities, and non-governmental organizations that have expertise in various aspects of real property management, such as commercial real estate, community planning and historic preservation.

Colocation among United States Postal Service Properties[37]

S. 1398 defines the term "postal property" to mean "any building owned by" the USPS. The bill exempts property "operated and maintained" by the USPS from the legislation's definition of "surplus property." The legislation does not, however, exempt USPS properties from its definition of "underutilized property." S. 1398 defines the USPS as a "federal agency" and therefore would require the USPS to carry out the "duties of federal agencies" enumerated in the legislation.

S. 1398 states that "each year, the Postmaster General [of the USPS] may ... identify a list of postal properties with space available for use by Federal agencies ... and submit the list" to the FRPC. Subsequently, the FRPC shall share this list with federal agencies, which will have 90 days to examine it and recommend colocation of their agencies into USPS properties. S. 1398 would allow the agencies and the USPS to conclude any colocation lease.

Existing USPS Real Property Authority

Congress has given USPS independent authority to acquire and dispose of its real estate as it deems proper.[38] Allowing USPS to make decisions over its real estate and property holdings has been viewed as integral to the concept of the USPS as encapsulated in the Postal Reorganization Act (PRA).[39] This 1970 statute replaced the Post Office Department with the USPS, and required it to be financially self-supporting. With the PRA, Congress relinquished a great deal of control over USPS's operations as that control had proven problematic.[40]

The PRA assigned USPS the "general duty" to "maintain an efficient system of collection, sorting, and delivery of the mail nationwide."[41] To carry out this obligation, Congress provided USPS with a number of powers to generate revenue and control its operational costs, including authority to "determine the need for post offices, postal and training facilities and equipment, and ... provide such offices, facilities, and equipment as it determines are needed."[42] This authority has helped USPS respond to shifts in population by expanding its presence in areas where the number of people and businesses was growing, and scaling back USPS's operational footprint in places where population were decreasing. This authority over its property and facilities also permits USPS to alter its logistical (mail-moving) network to accommodate mail volume changes and technological developments in mail processing.[43]

S. 1398 and Existing USPS Real Property Authorities

S. 1398's colocation provisions do not appear to reduce the USPS's long-standing authorities over lease-making. The legislation appears to intend to facilitate, not compel, the USPS to enter any colocation leasing agreements.

As noted above, however, the bill defines the USPS as a "federal agency" for the purposes of the legislation, which means the USPS would be obliged to carry out the 10 "duties" in the legislation. These duties, as described on page 8 in this report, include conducting annual property surveys, establishing property inventory controls and accountability measures, submitting to the FRPC reports on excess and underutilized USPS properties, and establishing goals to reduce excess and underutilized USPS properties.

Leasing

S. 1398 would require agencies with independent leasing authority to submit to the FRPC a list of all leases, and, for each lease, the dates the lease was executed and will expire; a description of the size of the property and its address; the tenant agency; the total annual rent; and the net present value of the lease over the life of the contract.[44]

Real Property Database

S. 1398 would require the GSA Administrator to establish and maintain a "single, comprehensive, and descriptive" database of all real property under the control of federal agencies. The database would include all of the information required in Section 629(9)(B) of the bill plus a list of real property disposals completed, including, for each property

- the date of disposal and disposal method used;
- the proceeds obtained;
- the amount of time required to dispose of the real property, including the date the property was designated as excess;
- the date on which the property is designated as surplus; and
- all of the costs associated with the disposal.

Once the database was operational, it would be made available—on request—to the Committee on Homeland Security and Governmental Affairs and the Committee on Environment and Public Works in the Senate, and the Committee on Government Oversight and Reform and the Committee on Transportation and Infrastructure in the House. In addition, the database must be accessible to the public at no cost within three years.

Expedited Disposal Program

Section 627 of S. 1398 would establish a five-year pilot program that would streamline the disposal process. The OMB Director would be permitted to authorize the expedited disposal of up to 200 surplus properties a year. The disposal may occur through transfer, sale, conveyance, or demolition, and priority would be given to properties with the highest fair market value and the greatest potential for disposal. GSA would be permitted to obligate funds to pay for the costs of identifying and preparing properties to be reported as excess—the first step in the disposal process. GSA would be repaid through the proceeds of any sale of surplus properties. Properties in the pilot program may be disposed of only if the proceeds would exceed the costs of disposal and are not less than 90% of fair market value.

All properties selected for the disposal program would be exempt from a range of provisions in existing laws, including statutory provisions that would require agencies to offer the properties for public benefit conveyance. Properties may be sold, transferred, or demolished, for example, without first being offered to aid the homeless or for other public purposes, as current law requires. Proceeds from the disposal of real property would be distributed as follows:

- 80% would be returned to the U.S. Treasury for debt reduction;
- the lesser of 18% or the share of proceeds otherwise authorized to be retained under law would be retained by the landholding agencies;

- not more than 2% would be used to fund homeless assistance grants (as described in Section 627 of the bill); and
- any remaining proceeds would be returned to the Treasury for deficit reduction.

Agencies would have two years to use the proceeds they received, but only after use of those funds had been authorized in annual appropriations acts, and only for real property management and disposal. If a surplus property in the pilot program was not disposed of within two years of being listed for sale, then it may be conveyed to state and local governments or qualified nonprofits, with the exception of properties not used for housing that have an area greater than 25,000 square feet or a fair market value in excess of $1 million.

If an agency fails to make a surplus property available for public sale within 18 months after the property was selected for the pilot program, that agency would not be permitted to increase the size of its civilian real property inventory unless the square footage of the proposed increase was offset through consolidation, colocation, or disposal of other space from that agency's inventory.

Homeless Assistance Grants

S. 1398 would permit the Secretary of the Department of Housing and Urban Development (HUD) to use funds made available from the disposal of pilot program properties to provide grants to eligible private nonprofit organizations. Eligible nonprofits must use any grant funds they receive to acquire or rehabilitate property in order to provide housing or shelter for the homeless. Grant recipients must also agree to use the property only for providing homeless services for at least 15 years.

H.R. 328: Excess Federal Building and Property Disposal Act of 2013

H.R. 328 was introduced January 22, 2013, and referred to the Committee on Oversight and Government Reform. The committee ordered it to be reported without amendment favorably on March 20, 2013. H.R. 328 also takes a broad approach to real property disposal reform. It includes provisions that would expedite the disposal of certain high-value properties, reduce the scope of the McKinney-Vento Act, and require landholding agencies to implement policies and practices that would reduce the number of unneeded properties in their portfolios. The bill would also require the establishment of a real property database that might enhance congressional oversight.

Expedited Disposal Program

The bill would establish a real property disposal program under which the GSA Administrator and the OMB Director, based on recommendations from landholding agencies, would identify the 15 federal properties that are excess or surplus and have the highest fair market value and the greatest potential to sell. Those properties would then bypass statutory transfer and conveyance requirements and be offered for sale immediately through public auction. Upon the sale of a property, the Administrator and Director would select another high-value property to take its place, thus maintaining a pool of 15 properties for sale under the program at all times. Properties subject to BRAC legislation, properties owned by the USPS, certain Indian and Native Eskimo properties, and properties the Administrator determined are suitable for use as a public park or recreation space would be excluded from the program. It appears that the bill would not exclude federal properties located outside the United States from the expedited disposal program.

The expedited disposal program under H.R. 328 would not permit the sale of properties in the program for less than fair market value or if the property would not generate revenue in excess of the costs of disposal. In addition, properties selected for the H.R. 328 expedited program would be exempt from a range of provisions in existing laws, including statutory provisions that would require agencies to screen the properties for homeless use and public benefit conveyance.

Under H.R. 328, proceeds generated by the disposal of properties under the program would be deposited into the Treasury's General Fund, with 2% of that amount authorized for homeless assistance grants as authorized in Section 625 of the bill. H.R. 328 would permit HUD to use the proceeds from the disposal of properties for grants to eligible private nonprofit organizations that aid the homeless.

Duties of the General Services Administration and Executive Agencies

H.R. 328 would require GSA to issue guidance on the development and implementation of agency real property plans, including recommendations for identifying excess properties, evaluating the costs and benefits associated with disposing of real property, and prioritizing disposal decisions based on agency missions and anticipated future holdings.

Executive agencies would be required to maintain adequate inventory controls and accountability systems, identify underutilized properties through ongoing surveys, report underutilized property to GSA, and transfer or dispose of excess property as promptly as possible. H.R. 328 would also require agencies to develop and implement a real property plan, identify and categorize all real property owned, leased or otherwise managed by the agency, and establish goals for reducing excess property in the agency's inventory. Finally, H.R. 328 would require agencies, "as far as practicable," to reassign underutilized property to another activity within the agency, transfer underutilized property to other federal agencies, and obtain underutilized properties from other federal agencies first before acquiring nonfederal property. The bill does not appear to exclude properties held by federal agencies in foreign countries.

The bill would require GSA to issue a report within three years of enactment that would detail the efforts of each agency to reduce its excess and surplus properties, and for each property disposed of, the date, method, and cost of the disposal, the proceeds obtained from disposition, and the amount of time required to complete the disposal.

Agency Retention of Proceeds

The cost of bringing a property to market would be paid out of proceeds generated from the sale, lease, or transfer of real properties that were not included in the expedited disposal program. The remaining amount—net proceeds—would be deposited into the real property account of the agency that had custody of the property at the time it was declared excess. H.R. 328 would require net proceeds to be authorized for expenditure in annual appropriations acts, and those funds, if appropriated back to the agency, may only be used for real property activities. Any net proceeds not expended would be used for deficit reduction.

Federal Real Property Database

H.R. 328 would require GSA to establish a database of all federal real property other than properties excluded for purposes of national security. The database would have to be accessible to Congress and the public, and it must include

- the location and size of each property;
- the relevance of each property to the agency's mission;
- the level of utilization of each property, including whether it was excess, surplus, underutilized, or unutilized, and the number of days each property was designated as such;
- the annual operating costs of each property; and
- the replacement value of each property.

Under H.R. 328, the database must also use a machine-readable format, and permit users to search, sort, and download data.

Sustainable Disposal of Property

H.R. 328 would require the head of each of each agency to divert at least 50% of construction and demolition materials and debris by the end of year 2015. While the legislation does not define "divert," this term typically refers to recycling or reusing materials that otherwise would be disposed of at a landfill.[45]

Streamlining the McKinney-Vento Homeless Assistance Act

Under H.R. 328 agencies would not have to put all of its properties through the screening process required by the McKinney-Vento Act. Specifically, the bill would exclude from homeless screening all properties that were located in an area for which the general public is denied access in the interest of national security. In addition, H.R. 328 would not eliminate the requirement that HUD publish a list of all surplus properties approved to assist the homeless in the *Federal Register* as current law requires. Instead, the bill would require this information to be published on a HUD or GSA website.

H.R. 695: Civilian Property Realignment Act of 2013

H.R. 695 was introduced on February 14, 2013, and referred to three committees: Transportation and Infrastructure, Oversight and Government Reform, and Rules. H.R. 695 would draw on the military base realignment and closure (BRAC) model of real property disposal by establishing an independent commission to assess agency portfolios and to recommend actions for reducing the government's inventory of unneeded and underutilized buildings. The bill may enhance congressional oversight because it would require the House and the Senate to approve of all disposal recommendations before agencies could implement them. Conversely, the bill may limit congressional oversight because the House and Senate would have only 45 days to review and vote on all of the recommendations, of which there may be hundreds.

Scope

H.R. 695 has a broad scope, applying to space owned and leased by all executive branch agencies and government corporations—not just properties that are excess or surplus. The bill would exclude some properties, such as those under the jurisdiction of the DOD or the Coast Guard, properties owned by the USPS, certain Indian and Native Alaskan properties, certain properties associated with land and water management programs, and properties located outside the United States that are operated or maintained by the Department of State or the United States Agency for International Development. The legislation would encompass most major real property asset management functions, collectively referred to as "realigning" actions—including the consolidation, reconfiguration, colocation, exchange, sale, redevelopment, and transfer of unneeded or underutilized properties.

Development of Recommendations

The first step in the process proposed by H.R. 695 would be for federal landholding agencies to develop their own recommendations for realigning their real property portfolios and for reducing operating and maintenance costs. Agencies would submit these recommendations to GSA and OMB not later than 120 days after the start of each fiscal year, along with specific data on each of the properties they own, lease, or otherwise control. The data would include the age and condition of the property, its operating costs, size in square feet (broken out by gross, rentable, and usable footage), number of federal employees and functions housed in the property, and the history of capital expenditures. The recommendations would include categorization of properties into those that can be sold, transferred, exchanged, consolidated, relocated, redeveloped, reconfigured, or otherwise disposed of so as to reduce the costs of operating and maintaining the federal real property portfolio.

The GSA Administrator and the OMB Director would also work together to develop criteria that they would use to determine which properties should be realigned and what type of realignment should be recommended (e.g., sale, consolidation, conveyance for public benefit) for each property. The bill specifies that nine "principles" must be taken into account when establishing the criteria; some of the supporting data may already be collected by agencies as they develop their asset management plans or meet existing reporting requirements, such as those for the FRPP.

- The extent to which federal buildings or facilities could be sold or redeveloped in a manner that would produce the best value.
- The extent to which the operating and maintenance costs would be reduced through the consolidation, colocation, and reconfiguring of space.
- The extent to which the utilization rate is being maximized and is consistent with nongovernment standards.
- The potential costs and savings over time.
- The extent to which leasing long-term space would be reduced.
- The extent to which a property aligns with the current mission of the agency.
- The extent to which there are opportunities to consolidate similar operations across or within agencies.

- The economic impact on existing communities in the vicinity of the property.
- The extent to which energy consumption specifically would be reduced.

The OMB Director would then conduct an independent analysis of agency recommendations and revise them, as deemed appropriate. The OMB Director would then submit the revised recommendations, along with the criteria, to a newly established Civilian Property Realignment Commission. The commission would be composed of nine members, each serving a six-year term. The chair would be appointed by the President, with the advice and consent of the Senate. The President would appoint the other eight members of the commission, but would also be required to consult with the Speaker of the House regarding the appointment of two members, the minority leader of the House regarding one member, the Senate majority leader regarding two members, and the minority leader of the Senate regarding one member. H.R. 695 would also require that the commission include members with expertise in commercial real estate and redevelopment, government management or operations, community development, or historic preservation. The commission would terminate after six years.

The commission would review the OMB Director's recommendations. As part of the review process, the commission would be required to develop an accounting system to help evaluate the costs and returns of various recommendations. In addition, if the commission chose to hold hearings on the recommendations, then the bill would require those hearings to be open to the public. The bill would also require the commission to include in its recommendations at least five federal properties not listed as excess or surplus but that have an estimated fair market value of at least $500 million, in total. H.R. 695 does not specify that all of the high-value assets must be sold, although the commission may recommend selling one or more of them.

While the commission "shall seek to develop consensus" in its recommendations, the report may include recommendations supported by only a majority of commission members. The commission would be required to submit its final recommendations to the President, and to establish a website and post its findings, conclusions, and recommendations on it. H.R. 695 would require GAO to publish a report on the recommendations, including a review of the methodology used to select properties for realignment.

Review by the President

The President would be required to review the commission's recommendations and submit, within 30 days of receiving them, a report to Congress that identifies which recommendations are approved, and which, if any, are not. If the President approves all of the commission's recommendations, then he must submit a copy of the recommendations to Congress along with a certification of his approval. If the President disapproves of some or all of the commission's recommendations, he would be required to submit a report to Congress and to the commission identifying the reasons for disapproval, and the commission would have 30 days to submit a revised list of recommendations to the President. If the President approves of all of the revised recommendations, he must submit a copy of the revised recommendations along with a certification of his approval to Congress. If the President does not submit a report within 30 days of the receipt of the commission's original or revised recommendations, then the process terminates for the year and agencies are not required to dispose of any properties under H.R. 695. In effect, the President would be able only to approve or reject a complete list of recommendations. He would not be able to amend the commission's recommendations himself before approving them.

Congressional Consideration of the Recommendations

After receiving the recommendations approved by the President, Congress would have 45 days to review them and debate their merits. Congress would be required to vote on a joint resolution of approval by the end of that period. As with the President, Congress would have the authority only to act on the entire list, not to approve or disapprove of individual recommendations. If no joint resolution of approval is passed within the 45-day time limit, or if the resolution is passed and the President vetoes it, then agencies would not be required to implement the recommendations.

Implementation

If a joint resolution of approval were enacted, agencies would be required to begin implementation not later than two years from the date the President transmitted the recommendations to Congress, and to complete implementation no later than six years from the same date. The GSA Administrator would be given authority to "take such necessary and proper actions, including the sale, conveyance, or exchange of civilian real property, as required to implement the Commission recommendations" as enacted. Other federal agencies must either use their existing authorities to implement the recommendations or work with GSA to do so. The Administrator would also have the authority to convey property for less than fair market value or for no consideration at all. This would appear to permit agencies to bypass steps in the existing disposal process. A property recommended for public sale, for example, may not have to go through the public benefit screening process. H.R. 695 would require the Secretary of HUD to evaluate "to the extent practicable" certain properties for homeless use as required under the McKinney-Vento Act. The provision would apply to properties identified for disposal in an enacted joint resolution of approval that were not more than 25,000 square feet or were valued at less than $5 million.

H.R. 695 would also expand the reporting requirements for all real property actions that exceed the prospectus threshold—the dollar amount established in 40 U.S.C. Section 3307 above which agencies must obtain approval from the House Transportation and Infrastructure Committee and the Senate Environment and Public Works Committee. The bill would require each prospectus to include a statement of whether the proposal was consistent with H.R. 695 and how life-cycle cost analysis was used to determine long-term costs, the life-cycle cost of a building, and "any increased design, construction, or acquisition costs identified" that are offset by lower long-term costs.

Funding

H.R. 695 would establish two accounts: a salaries and expense account to fund the commission's administrative and personnel costs, and an asset proceeds and space management fund (APSMF), which would be used to implement recommended actions. Both accounts would receive funds from appropriations—the bill authorizes a one-time appropriation of $20 million for the salaries and expenses account and a $62 million appropriation for the APSMF—but the APSMF would also receive the proceeds generated by the sale of properties pursuant to the commission's recommendations. The sales proceeds deposited in the APSMF account could only be used to cover the costs associated with implementing the commission's recommendations.

Leasing Authority

H.R. 695 would require most executive agencies seeking to acquire leased space to do so only by working through GSA. This restriction would not apply to VA properties or properties excluded for reasons of national security by the President. This requirement may facilitate oversight by consolidating leasing decisions with a single agency, although it is not clear whether this would restrict GSA's ability to delegate leasing authority to other agencies. If agencies were no longer able to use independent or delegated leasing authority, it could delay the acquisition of space needed to carry out their missions.

Life-Cycle Costs

H.R. 695 would require the Administrator to take a building's life-cycle cost into account when constructing or leasing a building. This requirement would apply only to buildings that had estimated construction costs of more than $1 million, the federal portion of the estimated construction or lease costs exceed 50% of the total costs, and, in the case of a lease, the property has more than 25,000 square feet. The bill would define "life-cycle cost" as the total sum of

- investment costs;
- capital costs;
- installation costs;
- energy costs;
- operating costs;
- maintenance costs; and
- replacement costs.

The bill would define "lifetime of a building"—the length of time over which the life-cycle costs would be calculated—to be 50 years or the period of time during which the building is projected to be utilized. The GSA Administrator, when submitting a prospectus to acquire space, would be required to include in the prospectus a statement of how the life-cycle cost analysis was used and whether the analysis identified potential costs that could be offset by lower long-term costs.

Comparison and Analysis of Key Provisions

Table 1, below, compares key provisions from each of the three proposals examined in this report—S. 1398, H.R. 328, and H.R. 695. An analytical discussion follows **Table 1**.

Expedited Disposal

Agencies have long argued that public benefits conveyance requirements, particularly those that require screening for homeless use, create an administrative burden that delays disposition and drives up maintenance costs. Savings, therefore, may be generated by permitting agencies to bypass screening requirements and move through the disposal process more quickly. Under H.R. 695, agencies would not be permitted to go beyond their existing authorities when disposing of properties as required by enacted recommendations, although GSA would be given the authority to "take such necessary and proper actions" to implement the commission's recommendations. In addition, the identification of specific properties for specific disposal or realigning actions may permit those properties to bypass statutory requirements that may otherwise have applied, such as those regarding public benefit conveyance. By contrast, H.R. 328 and S. 1398 explicitly exempt properties from public benefit conveyance requirements, but the exemptions under H.R. 328 would apply only to the 15 "high-value" properties that would be included in the program at any given time, while the exemptions under S. 1398 would apply to all properties recommended for disposal—as many as 200.

Final Disposal Recommendations

H.R. 695 proposes establishing a new Civilian Property Realignment Commission that would be responsible for the final list of recommendations to be considered by Congress. In addition, the bill would require the President to seek the consent of the Senate and to consult with leaders in both chambers, which could enable Congress to influence the composition of the commission. On the other hand, consultations with congressional leaders and Senate confirmation of the commission chair could slow down the development of recommendations if the nominations of several commission members of the nomination process are delayed.

S. 1398 would not create a new body to oversee the disposal process, but would instead utilize the existing Federal Real Property Council to develop asset management plans for each agency— plans that would include recommendations for disposal of underutilized properties. Membership on the FRPC would not be subject to congressional approval, but would ostensibly require that some of the most knowledgeable real property officials from each agency play a central role in improving real property management by developing government-wide asset management principles and policies, as well as by vetting and finalizing recommendations to the OMB Director regarding which properties should be disposed of and by what method. Under H.R. 328, agency heads would recommend properties for expedited disposal under the program that the bill would establish, but the OMB Director and the GSA Administrator would make the final selections.

Table 1. Comparison of Key Provisions of Select Real Property Proposals in the 113th Congress

	S. 1398	H.R. 328	H.R. 695
Expedited Disposal Provisions	Apply to not more than 200 surplus properties	Apply to 15 “high-value” properties	Apply to all recommended disposals
Final Disposal Recommendations Proposed	By OMB Director	By OMB Director in consultation with GSA	By Civilian Property Realignment Commission
New Congressional Actions Required for Disposal	None	None	Joint Resolution of Approval to permit implementation
Real Property Database and Reporting Requirements	GSA would establish public website; FRPC would submit in-depth report to OMB Director	GSA would establish public website and submit in-depth report to Congress	Commission would establish public website and have access to agency portfolio data

Congressional Action on Recommendations

H.R. 695 would require a 45-day timeframe for congressional action. Congress would have less than seven weeks to review all of the recommendations—of which there may be hundreds— which could reduce oversight of major real property actions. Consolidation projects, for example, are often complex, multi-year efforts, with long-term consequences for the agencies and communities involved, and for which Congress is asked to provide hundreds of millions, or even billions, of dollars. For this reason, Congress regularly holds hearings on major consolidation proposals. For example, the effort to consolidate the Department of Homeland Security at St. Elizabeth's in the District of Columbia (DC) is estimated to cost $3.26 billion and has been the subject of several congressional hearings.[46] The consequences of the consolidation are wide ranging, and include changing traffic patterns in Washington, DC, relocating thousands of employees, and ensuring historic preservation requirements are met. Similar issues have been raised regarding the consolidation of Food and Drug Administration headquarters, a project that has received hundreds of millions of dollars since FY2000. Congress may not feel it has sufficient time, under the proposed time constraints, to either approve or disapprove of the recommendations. S. 1398 and H.R. 328, on the other hand, would not require Congress to approve or disapprove a list of recommendations: both bills would use programs that are managed entirely by executive agencies.

Requiring Congress to approve or disapprove of the entire list of recommended actions could reduce conflict between various stakeholders interested in the properties in question. Some civilian agencies have found their disposal efforts complicated by the involvement of state and local governments, nonprofits, businesses, and community leaders with competing agendas. In 2002, for example, the USPS identified a number of "redundant, low-value" facilities that it sought to close in order to reduce its operating costs. As part of the facility closure process, USPS was required to formally announce its intention to close each facility and solicit comments from the community. USPS ultimately abandoned its plans to close many facilities it identified— including post offices that were underutilized, in poor condition, or not critical to serving their geographic areas—in part due to political pressure from stakeholders.[47] By moving the locus of decision making away from agencies and placing it in the hands of an independent commission, the amount of pressure that stakeholders exert on the process might be reduced.

Real Property Database and Reporting

As discussed earlier in this report, basic data on the federal real property portfolio—including information on how many excess and surplus properties each agency holds—are currently limited. H.R. 695 would require the commission to establish and maintain a public database with "relevant information" about the commission's recommendations. H.R. 695 would also require GAO to perform a detailed analysis of the recommendation and selection process, although no timeline is specified for the completion of the report. The commission, however, would have the authority to access all information pertaining to the recommendations, including detailed data on each property's age, condition, operating costs, size, and the number of employees housed at the property. The commission would also have access to other data that may be used by agencies when making their recommendations, such as the potential costs and savings of each realignment proposal. The commission itself would be required to post a report on its findings, conclusions, and recommendations on its own website, which may result in agency-level data being made public through the commission.

S. 1398 would require the FRPC to submit a report to the OMB Director that contains descriptive data similar to the report required under H.R. 695. S. 1398 would also require GSA to establish a descriptive database that must be available at no cost to the public, and it specifies that that the database must include certain data that may be of use to Congress. If the database includes all of the data currently stored in the Federal Real Property Profile (consistent with national security concerns), and allows users to search and sort the data, then it could be a useful oversight tool. However, if the data used in the new database are of the same quality as the data in the FRPP, it could decrease the utility of the data for making decisions or conducting analyses. This is true for all reporting and database requirements in the three bills examined in this report.

As noted, H.R. 328 would require GSA to establish and maintain a database of all federal real properties (other than those excluded for reasons of national security) that would be accessible to the public at no cost. The database would be required to include a wealth of descriptive information of each property, and it would permit users to search, sort, and download data. This approach would potentially provide the widest public access to federal real property data, and is the only proposal that would require online data to be searchable and downloadable—functions that transparency advocates believe are important tools for effective public oversight of federal spending.

CONCLUDING OBSERVATIONS

Each of the bills analyzed in this report would establish procedures for selecting federal properties to sell and for the distribution of sales proceeds. Generally, each of the bills would apply net proceeds towards further real property disposals and reducing the federal deficit or debt. It is not clear, however, that much revenue might be generated under each bill, given the lack of even the most basic data needed for analysis. For example, it is not known how many excess, surplus, and underutilized properties are held by each agency, how much it would cost to bring each property to market, and the estimated fair market value of individual properties. FRPP data show that sales have not generated significant net proceeds—the amount of revenue remaining after the costs of bringing the property to market are deducted—in recent years. For example, in FY2010, the government sold 466 properties that generated in $57 million in net proceeds, and in FY2009 the government sold 2,228 properties that generated $50 million in net proceeds.[48] The costs of bringing properties to market—whether they are due to environmental remediation, a backlog of needed repairs, or historic preservation requirements—or the undesirable location of unneeded properties are among reasons that so little profit is generated through sales. The proposed bills may increase sales revenue, however, by bringing properties to market that are in more desirable locations. H.R. 695, for example, would both require the Civilian Property Realignment Commission to recommend at least five properties that are *not* identified as excess or surplus—and therefore not subject to disposal requirements—but which have relatively high fair market value ($500 million). Similarly, under S. 1398 the Director of OMB would have the authority to require agencies to sell properties that are not excess or surplus. If agencies are holding properties that are valuable, and which they have not declared excess—the first step in the disposal process—then these bills may provide a mechanism by which those properties may be brought to the market and possibly generate greater net proceeds than sales have in recent years. H.R. 328 would limit the scope of its real property disposal pilot program to properties that are declared excess or surplus, but it might also increase sales revenue and net proceeds by bringing the 15 properties most likely to sell at a high market value to be auctioned. If agencies invest their real property funds in bringing these properties to market as soon as possible, then valuable properties which might otherwise have been conveyed or slowly moving through the screening process would be up for sale weeks, months, or even years sooner than under the current process.

FRPP data also show that the reduction of operating and maintenance costs has yielded greater annual savings to the government than net proceeds from sales have. In FY2010 the government reduced its annual operating costs by $274 million—four times the amount of net proceeds from sales that same year.[49] These figures do not include savings reported in the FRPP data that are the result of transferring properties between federal agencies, since the operating and maintenance costs have only been shifted from one agency to another, not eliminated.

There are underutilized and vacant properties, perhaps thousands of them, which agencies cannot dispose of because they lack the funding to make needed repairs. The total cost of these repairs government-wide is not known, but several agencies have reported repair backlogs in excess of $1 billion.[50] Public-private partnerships (PPPs) may be an option for funding some of these repairs. While PPPs may be structured in many different ways, they generally entail a contractual relationship between a non-federal entity—defined here as a private sector entity or state or local government—and a federal agency, in which the non-federal entity provides the capital to renovate or develop an underutilized property in return for a share of the revenue the improved property generates. Some agencies have the authority to enter into specific types of PPPs—the Department of Veterans Affairs can enter into enhanced use leases, for example—which has enabled them to generate positive cash flow from underutilized or vacant properties. Expanding PPP authorities, however, might come with risks to the government. Many agencies may lack sufficient expertise among their staff to negotiate PPP contracts effectively, and the result may be a contract that is not in the best interest of the government.

End Notes

[1] Federal Real Property Council, *FY2010 Federal Real Property Report: An Overview of the U.S. Federal Government's Real Property Assets*, September 2011, p. 13. Underutilized buildings are those that have a certain percentage of space that is not being used, generally calculated as a ratio of occupancy to design capacity.

[2] U.S. Government Accountability Office, *Federal Real Property: The Government Faces Challenges to Disposing of Unneeded Buildings*, GAO-11-370T, February 10, 2011, pp. 4-8.

[3] 40 U.S.C.§524(a).

[4] Ibid.

[5] The disposal provisions of General Service Administration's (GSA) real property regulations do not apply to agencies with independent authority to dispose of their own properties.

[6] 40 U.S.C. §102.

[7] Ibid.
[8] Ibid.
[9] 40 U.S.C. §545.
[10] There are benefits to these requirements as well, but they are not the focus of this memorandum.
[11] U.S. Government Accountability Office, *Federal Real Property: Progress Made in Reducing Unneeded Property, but VA Needs Better Information to Make Further Reductions*, GAO-08-939, September 2008, p. 39.
[12] U.S. Government Accountability Office, *Federal Real Property: Progress Made Toward Addressing Problems, but Underlying Obstacles Continue to Hamper Reform*, GAO-07-349, April 2007, p. 40.
[13] U.S. Government Accountability Office, *High-Risk Series: Federal Real Property*, GAO-03-122, January 2003, p. 41.
[14] 16 U.S.C. §470.
[15] U.S. Government Accountability Office, *Federal Real Property: DHS Has Made Progress, but Additional Actions Are Needed to Address Real Property Management and Security Challenges*, GAO-07-658, June 2007, p. 42.
[16] U.S. Government Accountability Office, *Federal Real Property: Progress Made in Reducing Unneeded Property, but VA Needs Better Information to Make Further Reductions*, GAO-08-939, September 2008, p. 5.
[17] U.S. Government Accountability Office, *Federal Real Property: Progress Made Toward Addressing Problems, but Underlying Obstacles Continue to Hamper Reform*, GAO-07-349, April 13, 2007, pp. 40-41.
[18] Ibid.
[19] Ibid.
[20] Ibid., p. 16.
[21] U.S. Government Accountability Office, *Federal Real Property: Progress Made in Reducing Unneeded Property, but VA Needs Better Information to Make Further Reductions*, GAO-08-939, September 2008, p. 5.
[22] U.S. Government Accountability Office, *High-Risk Series: Federal Real Property*, GAO-03-122, January 2003, p. 40.
[23] There is no government-wide real property guidance for addressing stakeholder conflicts.
[24] Executive Order 13327, "Federal Real Property Asset Management," 69 *Federal Register* 5897, February 4, 2004. According to the provisions of E.O. 13327, only the 24 agencies listed in 31 U.S.C. 901(b)(1) and (b)(2), which are subject to the Chief Financial Officers Act, are required to report real property data to GSA. Those agencies are the Departments of Agriculture, Commerce, Defense, Education, Energy, Health and Human Services, Homeland Security, Housing and Urban Development, the Interior, Justice, Labor, State, Transportation, the Treasury, and Veterans Affairs; the Environmental Protection Agency; General Services Administration; National Aeronautics and Space Administration; National Science Foundation; Nuclear Regulatory Commission; Office of Personnel Management; Small Business Administration; Social Security Administration; and United States Agency for International Development.
[25] The annual real property summary reports may be found on GSA's Federal Real Property Report Library website, at http://www.gsa.gov/Portal/gsa/ep/contentView.do?contentType =GSA_BASIC&contentId=23962.
[26] U.S. Government Accountability Office, *Federal Real Property: An Update on High-Risk Issues*, GAO-09-801, July 15, 2009, p. 10.
[27] U.S. Government Accountability Office, *Federal Real Property: National Strategy and Better Data Needed to Improve Management of Excess and Underutilized Property*, GAO-12-645, June 2012, p. 1.
[28] Ibid.

[29] Federal Real Property Council, *FY2007 Federal Real Property Report: An Overview of the U.S. Federal Government's Real Property Assets*, May 2008, p. 33. Federal Real Property Council, *FY2008 Federal Real Property Report: An Overview of the U.S. Federal Government's Real Property Assets*, August 2009, p. 24. Federal Real Property Council, *FY2010 Federal Real Property Report: An Overview of the U.S. Federal Government's Real Property Assets*, September 2011, p. 13.

[30] Federal Real Property Council, *FY2010 Federal Real Property Report: An Overview of the U.S. Federal Government's Real Property Assets*, September 2010, p. 13.

[31] U.S. General Services Administration, Office of Governmentwide Policy, *FY2008 Federal Real Property Report*, August 2009, p. 30.

[32] U.S. Government Accountability Office, *Federal Real Property: Overreliance on Costly Leasing Contributed to High-Risk Designation,* GAO-11-879T, August 4, 2011, p. 2.

[33] Ibid.

[34] Federal Real Property Council, *FY2010 Federal Real Property Report: An Overview of the U.S. Federal Government's Real Property Assets*, September 2011, p. 4.

[35] Securities and Exchange Commission, Office of Inspector General, *Report of Investigation: Improper Actions Relating to the Leasing of Office Space*, May 16, 2011, p. 2.

[36] Ibid.

[37] This section authored by Kevin R. Kosar (x7-3968).

[38] Thus, for example, Congress exempted USPS from the fe4deral property disposition statutes (40 U.S.C. §101).

[39] P.L. 91-375; 84 Stat. 725 (39 U.S.C. §101 et seq.).

[40] Congressional directives often inhibited the Old Post Office Department from controlling its operating costs and increasing its prices.

[41] 39 U.S.C. §403(b).

[42] 39 U.S.C. §404(a)(13).

[43] Over the past century, the proportion of mail sorted by postal workers has greatly declined with the advent of machinery that can read the addresses on letters and parcels and sort them according to their destination.

[44] Net present value is a financial calculation that accounts for the time value of money by determining the present value of future savings minus up-front investment costs over a specific period of time.

[45] As part of its Strategic Sustainability Performance Plan (SSPP), for example, GSA has set a goal of diverting "50% of its nonhazardous solid waste and construction and demolition debris from landfills through recycling, re-use of materials, composting organic waste, and thermal treatment." GSA's entire SSPP may be found at http://www.gsa.gov/portal/content/187149.

[46] U.S. Government Accountability Office, *Federal Real Property: DHS has Made Progress, but Additional Actions are Needed to Address Real Property Management and Security Concerns,* GAO-07-658, June 2007, p. 4. U.S. Congress, House Committee on Transportation and Infrastructure, *Subcommittee on Coast Guard and Maritime Transportation, Review and Status of the Multi-Billion Dollar Department of Homeland Security Relocation Project in Washington, D.C., and its Impact on the U.S. Coast Guard,* 112th Cong., 1st sess., September 23, 2011 (Washington, DC: GPO, 2011). U.S. Congress, House Committee on Appropriations, Subcommittee on Homeland Security, *St. Elizabeths and the Department of Homeland Security Consolidation*, 111th Cong., 2nd sess., March 25, 2010 (Washington, DC: GPO, 2010).

[47] U.S. Government Accountability Office, *Federal Real Property: An Update on High-Risk Issues*, GAO-09-801, July 15, 2009, p. 15.

[48] Federal Real Property Council, *FY2010 Federal Real Property Report: An Overview of the U.S. Federal Government's Real Property Assets*, September 2011, p. 13. Federal Real Property Council, *FY2009 Federal Real Property Report: An Overview of the U.S. Federal Government's Real Property Assets*, September 2010, p. 8.

[49] Ibid.

[50] U.S Government Accountability Office, *Federal Real Property: Government's Fiscal Exposure from Repair and Maintenance Backlogs Is Unclear*, GAO-09-10, October 2008, p. 2.

In: Federal Disposal Responsibilities
Editor: Tara C. Waller
ISBN: 978-1-63321-217-6

Chapter 2

MARITIME ADMINISTRATION: SHIP DISPOSAL PROGRAM NEEDS IMPROVED COMMUNICATIONS AND UPDATED STRATEGIC PLAN*

United States Government Accountability Office

WHY GAO DID THIS STUDY

Timely and proper disposal of obsolete ships in the National Defense Reserve Fleet—older ships designated for use in national emergencies—is critical to protecting the environment. Because these ships often contain hazardous materials, members of Congress and others have raised issues about the environmental concerns. As part of the Department of Transportation, MARAD's Ship Disposal Program serves as the federal government's agent for competing and awarding contracts for recycling the ships' materials. Congress has required MARAD's ship disposal program to award ship recycling contracts to qualified ship recycling facilities on the basis of best value. The Coast Guard and Maritime Transportation Act of 2012 mandated that GAO review MARAD's source selection procedures and practices used to award ship recycling contracts.

* This is an edited, reformatted and augmented version of the United States Government Accountability Office publication, GAO-14-223, dated February 2014.

In this report, GAO assessed MARAD's (1) source selection process; (2) communication strategy with ship recycling contractors; and (3) long-term ship disposal strategy. To complete this work, GAO reviewed and analyzed documentation on MARAD's qualification process, source selection procedures, and strategies; and interviewed MARAD and all of its qualified ship recycling contractors.

What GAO Recommends

GAO is recommending that the Department of Transportation improve its communication strategy and update its strategic plan. The Department agreed with the facts but did not take a position on our recommendations.

What GAO Found

The Maritime Administration (MARAD) uses a two-step source selection process, first by qualifying contractors and then awarding contracts for ship recycling services based on best value, consistent with the Federal Acquisition Regulation (FAR). In the first step, MARAD qualifies contractors' ship recycling facilities. The qualification process involves evaluating ship recycling facilities' proposals based on multiple criteria, including how a facility plans to dismantle ships and the extent to which the contractor—including its ability to meet local, state, and federal regulations—supports that effort. For the second step, MARAD awards ship recycling contracts for specific ships using a best value source selection process. The best value source selection process allows the government to accept an offer other than the best-priced offer, considering both price and non- price factors, that provides the greatest overall benefit to the government. MARAD considers three evaluation criteria—price, schedule and capacity, and past performance.

MARAD has made some efforts over the last year to clarify certain elements of its source selection process; however, MARAD could strengthen its communication strategy with its contractors. All of the qualified contractors GAO spoke with were confused about MARAD's source selection process—including how MARAD uses past performance to evaluate contractors' offers. MARAD has made an effort to clarify its past performance criterion by further explaining what is considered in its most recent solicitation. However other concerns remain. For example, some contractors expressed concern as to whether changes to their facility were approved by MARAD. GAO's standards for internal controls state that management should ensure adequate means of communicating with external stakeholders that may have a significant impact on the agency achieving its goals. Improving its communication strategy with its contractors could help MARAD maximize the transparency of its source selection process.

In 2006, MARAD issued a comprehensive management plan, or strategic plan, that outlined short- and long-term strategies for the disposal of MARAD's obsolete ships; however several key elements are now outdated or no longer applicable. According to program officials, MARAD was required to provide regular reports to Congress for several years, the last of which was in March 2011, on the progress made to address the backlog of obsolete ships. However, these reports did not provide the strategic short- and long-term direction for the program. Further, they indicated that the principles of the plan remain relevant; but the 2006 strategy does not take into account current market conditions, goals, and external risks. For example, concerns about maintaining the supplier base are not addressed in the plan. MARAD wants to maintain a supplier base to ensure competition for future ship recycling contracts, but has not fully considered risks and options to address this pending issue. Competition is a cornerstone in federal contracting and a critical tool for achieving the best return on the government's investment. An updated strategic plan that reflects the current external environment and risks could better position MARAD to identify future challenges and opportunities to help ensure the long-term participation and competition of the industrial base for ship disposal.

February 12, 2014

Congressional Committees

As part of the Department of Transportation, the Maritime Administration's (MARAD) Ship Disposal Program serves as the federal government's agent for competing and awarding contracts for disposal of obsolete ships in the National Defense Reserve Fleet (NDRF) and some of those from other federal agencies weighing 1,500 gross tons or more.[1] Many of these ships contain hazardous materials such as asbestos, lead, polychlorinated biphenyls, and residual oils and fuels. MARAD must comply with certain federal statutes that affect ship disposal including those related to the handling of pollutants and hazardous waste. However, members of Congress and environmental groups have raised issues about the potential environmental hazards—such as costly oil spills or other environmental threats—posed by these ships and the need for the timely and proper disposal of them to protect the environment. MARAD's disposal prioritization is based on ship material condition and a "worst first" disposal policy for obsolete ships in the NDRF. From fiscal year 2001 through fiscal year 2013, 197 ships have been removed from MARAD's fleets for disposal. As of October 2013, 27 obsolete ships remain in the fleet, while about three to five additional ships are declared obsolete each year.

MARAD can dispose of obsolete ships through sales recycling contracts—in which the government sells the ship to the contractor and the contractor dismantles the ship and sells the scrap for profit. Prior to 2000, MARAD sold ships to the contractor with the highest price for the ship. By the late 1990s, the market price for scrap metal decreased, which contributed to a decline in contractors' interest in purchasing ships. As a result, MARAD experienced a backlog of obsolete ships for disposal. Beginning in fiscal year 2001, Congress expanded MARAD's ship disposal authority to include the option to pay for recycling services through contracts for services—through which the government pays the contractor to dismantle and dispose of the ship's material.[2] Congress has also required MARAD, for both sales contracts and contracts for services, to select awardees on the basis of best value, consistent with the Federal Acquisition Regulation (FAR), considering, among other things, least cost to the government, timeliness, worker safety, and environmental concerns. When awarding these contracts on the basis of best value, MARAD selects a contractor's offer that, in the government's estimation, provides the greatest overall benefit in response to the requirement, considering both price and non-price factors.[3] At times, this may result in an award that is not the best price.[4] MARAD started using best value source selection procedures for ship recycling contracts beginning in 2002.

The Coast Guard and Maritime Transportation Act of 2012 mandated that GAO conduct an assessment of the source selection procedures and practices used to award MARAD's NDRF ship recycling contracts, including MARAD's processes, procedures, and practices for the qualification of ship recycling facilities, determining whether MARAD's contract source selection procedures and practices are consistent with the law and best practices associated with making source selection decisions, as well as any other aspect we deem appropriate to review.[5] This report assesses MARAD's (1) source selection process; (2) communication strategy with ship recycling contractors; and (3) long-term ship disposal strategy.

To determine MARAD's source selection process for ship recycling, including facility qualification and best value process for contract awards, and the extent to which MARAD communicates that process with contractors, we identified and reviewed federal laws and regulations associated with MARAD's source selection processes, including FAR regulations on simplified acquisition procedures and best value determinations, and compared these to MARAD's processes. We reviewed and analyzed the cognizant facility qualification documents, including the proposals submitted to MARAD. Further, to identify how MARAD made best value determinations, we identified and reviewed a random, non-representative sample of source selection documents from fiscal year 2010 through fiscal year 2013 to identify the steps and analysis conducted by MARAD to select a contractor for award. We selected contracts awarded from fiscal year 2010 to 2013 to include both sales contracts and contracts for services in our sample. We interviewed MARAD Ship Disposal Program officials responsible for source selection and all six qualified contractors and one contractor going through qualification to determine the facility qualification process and how MARAD communicates with contractors.[6] We also compared MARAD's communication strategy to criteria in *Standards for Internal Control in the Federal Government* related to effective communications.[7]

To assess MARAD's long-term disposal strategy, we reviewed MARAD's documentation related to its ship disposal strategies. We reviewed Standards for *Internal Control in the Federal Government* related to strategic planning, and assessed MARAD's actions against these standards. We also interviewed MARAD officials to identify MARAD's program goals and long term strategies related to ship disposal.

We conducted this performance audit from May 2013 to February 2014 in accordance with generally accepted government auditing standards. Those standards require that we plan and perform the audit to obtain sufficient, appropriate evidence to provide a reasonable basis for our findings and conclusions based on our audit objectives. We believe that the evidence obtained provides a reasonable basis for our findings and conclusions based on our audit objectives.

BACKGROUND

MARAD's Office of Ship Operations maintains the NDRF as a reserve of ships for defense and national emergencies. When ships are no longer considered useful for defense or aid missions, the Office of Ship Disposal, which reports to the Associate Administrator for National Security, arranges for their responsible disposal. NDRF ships are located at three sites: the James River Reserve Fleet at Fort Eustis, Virginia; the Beaumont Reserve Fleet in Beaumont, Texas; and the Suisun Bay Reserve Fleet at Benicia, California. In addition, as established in a Memorandum of Agreement with the Navy in 2011, MARAD disposes of certain non-combatant ships from the Naval Inactive Ship Maintenance Facilities in Pearl Harbor, Hawaii and in Philadelphia, Pennsylvania. Naval Inactive Ship Maintenance Facilities are used to hold decommissioned U.S. Navy and auxiliary ships pending determination of their disposal.

MARAD's Ship Disposal Program's mission is to dispose of obsolete NDRF ships through the most expedient, best value, and environmentally safe disposal methods. The program's approach is to remove all ships that present the highest risk to the environment as soon as possible, and to have disposal alternatives and the necessary funding in place to ensure that obsolete ships can be disposed of at a faster rate than ships in the NDRF are designated as obsolete. MARAD's Ship Disposal Program is authorized to consider alternative ship disposal methods in addition to recycling obsolete NDRF ships, including, for example, artificial reefing and donation. MARAD has made limited use of these methods because of a number of environmental, financial, and legal barriers. According to MARAD, recycling is the most expedient, environmentally sound, and cost-effective method of disposal available.

After MARAD determines that a ship is ready for recycling, it first offers the ship for sale to qualified contractors through sales contracts. If it receives no offers to purchase the ship, MARAD will then procure services to dispose of the ship through a contract for services—in which the government pays the contractor to recycle the ship's material. When acquiring services through a contract, MARAD uses simplified acquisition procedures, as defined in the FAR, which allow for simplified processes to award contracts under a certain dollar threshold.[8] MARAD is generally not required to follow the procedures delineated in the FAR when awarding sales contracts because the government is not acquiring supplies or services with appropriated funds. However,

MARAD officials explained that because they are required to adhere to the FAR when awarding contracts for services, to be consistent, they adopted similar procedures for sales contracts.

In the past two years, MARAD has disposed of the ships exclusively through domestic recycling using sales contracts. See appendix II for more information on the ship disposal options used since 2001. In its most recent announcement in November 2013, however, three ships were announced as ready for recycling and MARAD received no sales offers. As a result, MARAD plans to procure the services to recycle the ships through contracts for services. MARAD officials explained that, while they cannot know all the factors the contractors considered when deciding whether or not to submit a sales offer on a ship, the specific ships in this announcement may not have looked profitable to the contractors for a variety of reasons. One ship was tested and proven to have polychlorinated biphenyls, which must be removed through a very costly process before the steel can be sold to recyclers. The other two ships were small, and MARAD officials noted that contractors may have estimated that the costs of towing the ship to the facility plus the labor were not likely to be recovered in the sale of the steel.

MARAD's Two-Step Source Selection Process Uses a Best Value Process Consistent with the Federal Acquisition Regulation

Since 2005, MARAD has been using a two-step source selection process for ship recycling. The first step is to qualify the facilities of ship recycling contractors that respond to MARAD's solicitation for ship disposal, thereby creating a pool of qualified facilities for contracts for either sales or services. In this step, MARAD determines if a ship recycling facility is qualified by evaluating the facility against multiple criteria as outlined in the solicitation. The second step is to award ship recycling contracts for specific ships to qualified contractors. Qualified contractors submit offers in response to MARAD's announcements—which inform the contractors about the location, names, and work to be completed on multiple ships available for recycling. MARAD then evaluates the offers and determines which offer is the best value for the government, consistent with the FAR, using price and non-price factors such as schedule, capacity, and past performance.

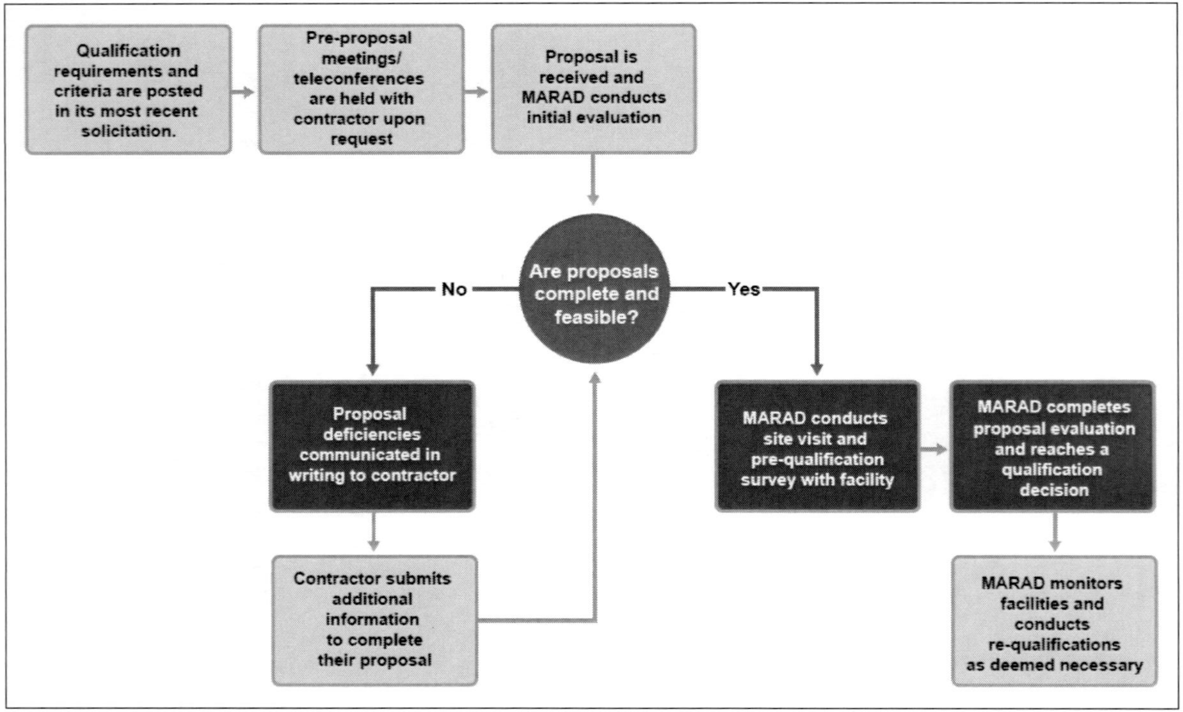

Source: GAO analysis of MARAD information.
See Appendix III for Additional Information Regarding Figure 1.

Figure 1. MARAD's Qualification Process of Ship Disposal Facilities.

MARAD's Qualification Process Is Based on Multiple Criteria and Varies in the Time It Takes to Complete

The first step for a contractor interested in having their facility qualified is to submit a proposal in response to MARAD's solicitation. MARAD's process for determining if a ship recycling facility is qualified consists of evaluating the facility's proposal against multiple criteria as outlined in MARAD's solicitation for ship disposal. These criteria include considering how a contractor plans to dismantle ships and the extent that the site is capable to support that effort. Beginning in 2005, MARAD began using an open solicitation process—in which interested vendors can respond at any time as opposed to by a specific date.[9] Periodically, MARAD will update and reissue the solicitation, as it did in fiscal year 2009 and again in fiscal year 2014. When the solicitation was updated in fiscal year 2014, previously-qualified facilities needed to re-qualify based on any new criteria incorporated into the updated solicitation.

Once the interested recycling contractor submits a proposal, MARAD follows the steps indicated in figure 1 to determine if the ship recycling facility is qualified.

MARAD reviews the proposal—including a technical compliance plan that addresses performance capability and ability to comply with all applicable local, state, and federal environmental, safety, and health laws and regulations—for completeness in its initial evaluation. Once MARAD determines that the proposal is complete, it begins evaluating the proposal against multiple criteria including:

- feasibility and likelihood of success of the engineering, technical, and management approach for disposal of obsolete ships and associated risks;
- productivity of the proposed approach including schedule and cost assessments;
- type and sufficiency of proposed methods, processes, and procedures; and
- depth, relevance, and currency of requisite experience of the facility.

The length of the qualification process—which has been as short as a few months or as long as five years—varies for several reasons. One reason for the variation, according to MARAD officials, is that the initial proposals are rarely complete. This part of the process is iterative and MARAD officials stated that they work closely with individual contractors to obtain the necessary information to enable the completion of the evaluation process. In its fiscal year 2014 solicitation, MARAD's stated goal is to complete the initial evaluation of proposals within eight weeks from the date the proposals are submitted. This timeline, according to officials, is based on the availability of MARAD's resources. However, MARAD officials noted that the contractor's responsiveness to requests for additional explanation or information is out of their control and can impact their ability to move forward with the qualification process. In some cases, a contractor may opt to not continue with the qualification process.

Another reason for the different timetables for qualification, according to MARAD officials, is that each facility is unique and faces individual challenges. MARAD officials explained that it is incumbent upon the contractor to identify all of the local, state, and federal laws and regulations that apply to ship recycling for the location where the facility is located, and to comply with those laws. For example, a facility in California took over five years to qualify because, according to MARAD officials, it had to address strict state and local environmental laws and regulations and experienced significant delays in obtaining necessary permits from state environmental regulators. Further, the proposed location had not been used for many years, and MARAD needed additional assurances to ensure qualification criteria were met.[10] When a new process is introduced by a ship recycling facility, qualification may take more time. For instance, a contractor recently proposed a new method for ship recycling that involves dismantling the ship while afloat. MARAD considers the procedures for maintaining the stability of the ship; environmental abatement of hazardous materials, such as asbestos and paint disposal; worker safety; and health considerations under this new approach before making a qualification decision.

As shown in figure 2, MARAD currently has six qualified recycling facilities, represented by four contractors, in two states: Louisiana and Texas. As of October 2013, there were 27 MARAD and two Navy-owned ships awaiting disposal.

After a facility is deemed qualified, MARAD conducts regular oversight, to include bi-weekly teleconferences and quarterly visits to the facilities. During its visits, MARAD checks the facility's compliance processes and issues deficiency notices if corrective action is needed. In addition, MARAD conducts oversight of the environmental protection and worker safety procedures for the facility. If there is a deficiency found, MARAD refers the issue to the appropriate federal agency, such as the Environmental Protection Agency or Occupational Safety and Health Administration. Further, through its announcement process, MARAD requires existing qualified contractors to report changes made to their ship recycling facilities. These changes could range from a minor change to one part of its recycling process, such as a change in paint removal process, or a significant change such as when a facility is changing ownership.

MARAD's Best Value Process for Source Selection of Its Ship Recycling Contracts is Consistent with the Federal Acquisition Regulation

After becoming qualified, a contractor's facility is eligible to compete for ship recycling contracts, and does so by providing offers in response to MARAD's announcements on specific ships. Approximately three to four announcements are issued per year, with three to six ships offered per announcement. MARAD's announcement includes the location and names of the ships that MARAD would like recycled, and the work to be accomplished for each ship. According to MARAD officials, including multiple ships in each announcement optimizes competition and minimizes costs.

To decide which facility should receive contracts for the ships in the announcement, MARAD uses a source selection process to determine which offer provides the best value to the government. This process allows the government to accept an offer other than the best priced offer, considering both price and non-price factors. MARAD uses the following three evaluation criteria to determine which offers provide the best value for the government:

- price,
- schedule and capacity, and
- past performance.

Because it is using simplified acquisition procedures, MARAD is not required to disclose the relative importance assigned to each evaluation factor.[11] For example, MARAD is not required to state in the announcement whether price is more, less, or equal in importance to past performance.

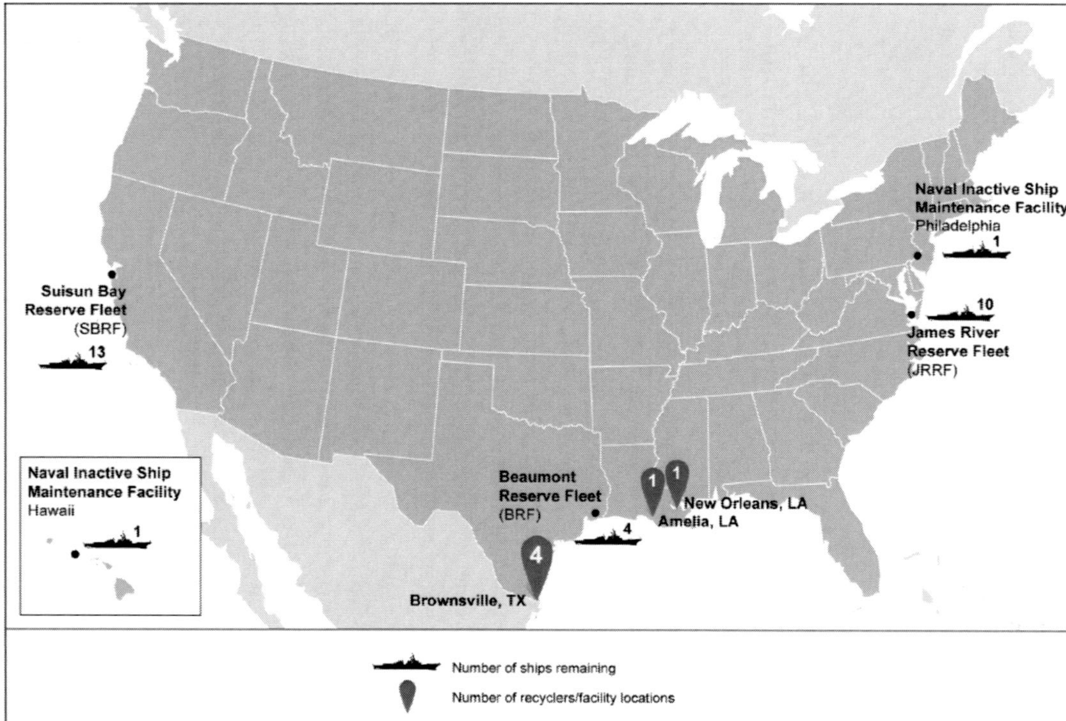

Source: GAO analysis of MARAD information.

Figure 2. Locations of National Defense Reserve Fleets and Number of Ships, Naval Inactive Ship Maintenance Facilities and Number of Ships, and Qualified Facilities as of October 2013.

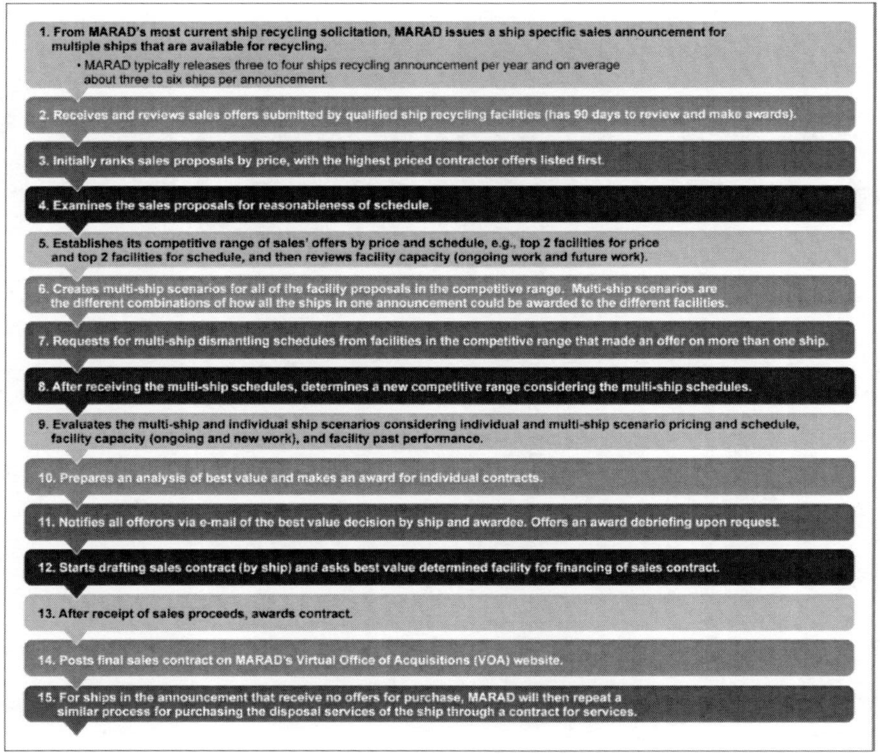

Source: GAO analysis of MARAD information.

Figure 3. Maritime Administration's (MARAD) Source Selection Process for Sales Contracts for Ship Recycling, Fiscal Year 2013.

Although not required to state the relative importance of the evaluation criteria, MARAD's source selection process generally addresses the criteria in a specific order. Initially, MARAD ranks contractors' sales offers on individual ships by highest to lowest price, then by shortest to longest schedule. MARAD then determines a competitive range—in other words, a subset of all the offers that represents the most highly rated offers. Once the competitive range is determined, MARAD requests multi-ship recycling schedules from contractors with competitive offers on more than one ship to gain insight into what the schedule would be if the facility wins more than one ship in the announcement. For example, if a contractor made offers on two ships that were determined to be in the competitive range, MARAD would ask the contractor to provide a revised schedule showing how many days it would take the contractor's facility to recycle both ships. MARAD then reviews capacity at each facility in the competitive range. Capacity, according to MARAD, refers to the facility's ability to take possession of and actively work on the ship(s) at its facility, which reflects current and future ship recycling work from all sources at the recycling facility. Finally, MARAD considers past performance, which can include timeliness on past contracts, a change in management, or the amount of time since the facility last recycled a MARAD ship. Figure 3 describes in more detail the steps MARAD uses in its best value source selection for ship recycling contracts.[12]

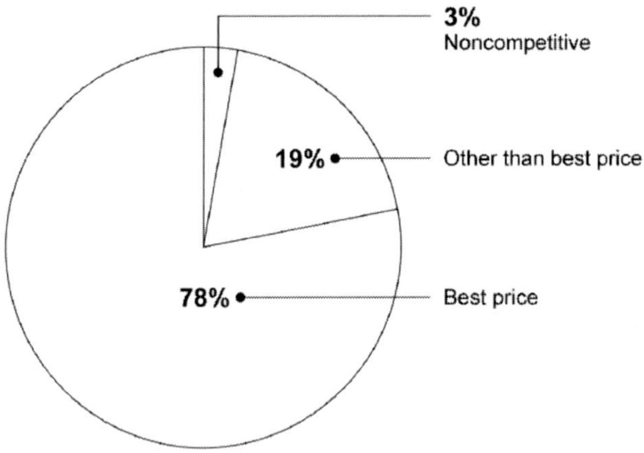

Source: GAO analysis of MARAD information.

Figure 4. Percentage of the Maritime Administration's (MARAD) Ship Recycling Contracts Awarded to the Best Price Offer, Other than Best Price Offer, and Non-Competitively, Fiscal Years 2009 through 2013.

As indicated in figure 4, most contracts were awarded to the contractor with the best price offer.

For fiscal year 2009 through fiscal year 2013, about 78 percent of MARAD's contract awards for ship recycling, including both sales contracts and contracts for services, were awarded to the contractor that offered the best price. Half of all contracts were awarded to the contractor that offered both the best price and shortest schedule. Another 19 percent of the awardees were selected based on factors other than the best price, and 3 percent were awarded non-competitively, which means these awards were not subject to MARAD's typical source selection process. For example, one contract was awarded non-competitively because it was in the best interest of the government to support an increased industrial base of ship recycling facilities on the west coast.

MARAD Has Taken Steps to Clarify Its Source Selection Process, but Could Strengthen Its Communication Strategy

MARAD has made some efforts over the last year to clarify certain parts of its source selection process; however, during our review, contractors told us that they do not fully understand how it works. Almost all of the contractors were unsure as to how MARAD determines best value under sales contracts. Most of the contractors told us that they did not always understand why they lost a ship recycling contract. Three contractors noted it was particularly confusing when they offered the best price with what they believed to be a reasonable schedule compared to other contractors, yet still lost the contract. A couple admitted that there were instances where they did not understand why they won a contract.

One consistent source of confusion was that they did not know how MARAD values past performance when awarding contracts for ship recycling—both in how past performance affected the chance to be awarded a future contract and in what factors were considered. For example, they did not know if environmental infractions or late performance on previous contracts played a role in the ability to win future MARAD ship recycling contracts. MARAD, for its part, explained to us that past performance has rarely affected a contractor's ability to win a contract for one ship in an announcement, but has had a greater affect when MARAD considers the award of more than one ship to the contractor. In addition, in our review of contract documentation, we found several instances where MARAD considered past performance in combination with capacity. For example, a contractor may have had schedule delays on prior ships which were then exacerbated when the contractor was managing the recycling schedule for multiple ships. How MARAD considers the multiple factors when making a best value award varies depending on the specific circumstances of the procurement. As a result, the basis for each award may be unique, which may contribute to the contractor's confusion on how MARAD is making these decisions. To clarify past performance, in the fiscal year 2014 solicitation, MARAD has elaborated on what will be evaluated within the past performance criterion—stating that it includes the contractor's history of reasonable and cooperative behavior, record of integrity, and business relationships with the customer. Although a positive step, lack of communication with contractors regarding effects of their past performance, if any, on their future offers remains a source of confusion.

Further, MARAD recently improved its contractor debriefing process by adding two more narrative sections to its pre-debriefing form, one on capacity and one on best value. Under the simplified acquisition process, if a contractor requests information on an award that was based on factors other than price alone, MARAD is to provide a brief explanation of the basis for the contract award decision. Although not required to do so, about two years ago, MARAD decided to offer more formal debriefings to convey this information. The debriefings are intended to help contractors understand how their offer compared to the winning offer and to help improve offers for the next time. Once a facility contractor requests a debriefing, MARAD creates a pre-debriefing report that compares the contractor's offer—price and schedule—to the successful contractor's price and schedule. The pre-debriefing report is then sent to the requesting contractor with an invitation to participate in a debriefing meeting. MARAD told us that all contractors have been informed of the availability, upon request, of debriefing after a contract award. However, according to MARAD, after the last four ship recycling announcements, only three of the six contractors that bid on the ships asked for a debriefing.

These three contractors stated that MARAD's debriefing process did not help them understand why they lost the contract or how MARAD considered price and non-price factors in its best value determinations. MARAD revised the debriefing form to include the additional narrative sections; however, it has not informed contractors of this change. Contractors may not be aware that the additional information is available unless they request a debriefing. MARAD officials told us that they plan to inform contractors of the new debriefing form through an amendment to the solicitation. If MARAD communicates the additions to the debriefing form in an open and transparent way, facility contractors may seek to obtain information in order to improve their offers for the next announcement, which ultimately could lead to better results for the government.

Contractors that we spoke with also expressed confusion about other parts of MARAD's source selection process, including its qualification process. For example, four contractors said that they notified MARAD of a change to an element of their recycling process at a facility, and that MARAD acknowledged receipt of the notification but did not indicate whether the change was approved or not approved. These contractors were confused about whether certain changes made would affect their eligibility to compete for a contract. In one instance, a contractor told us they informed MARAD about newly-installed ship cutting pads and some dredging they had completed at their facility, but they heard no response from MARAD. According to the contractor, a MARAD official only recognized the changes during a site visit to the facility.

In response to these concerns, MARAD officials explained that they are always available to hear questions from contractors should any questions arise, and contractors should feel comfortable contacting them with any questions. *Standards for Internal Control in the Federal Government* state that management should ensure there are adequate means of communicating with external stakeholders that may have a significant impact on the agency achieving its goals.[13] Although MARAD has improved some aspects of communication with facility contractors, it could strengthen its communication strategy. For example, one method to improve communication could involve an annual industry day or conference where all contractors can ask questions and request clarification on any part of MARAD's source selection process or other aspects of ship disposal resulting in all contractors hearing the same responses at the same time. Regular individual meetings might also be scheduled with contractors where their specific concerns are discussed. Improving its communication strategy with its contractors could help MARAD maximize the transparency of its source selection process.

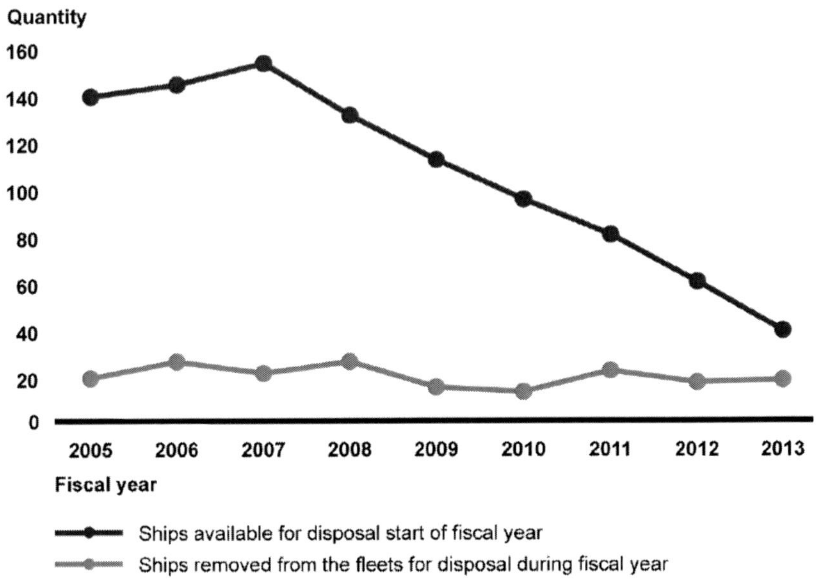

Source: GAO analysis of MARAD information.

Figure 5. Number of Ships in Fleet that were Available for Disposal at the Start of the Fiscal Year and Removed from the Fleet during the Fiscal Year, Fiscal Years 2005 through 2013.

MARAD'S STRATEGIC PLAN DOES NOT REFLECT CURRENT CONDITIONS

In our 2005 report, we found MARAD had not developed a comprehensive management approach that could address the myriad of environmental, legal, and regulatory challenges that the program faces. As a result, we recommended that it develop a comprehensive approach to manage its ship disposal program.[14] In 2006, MARAD issued a comprehensive management plan, or strategic plan, which outlined short and long-term strategies for the disposal of the agency's obsolete ships; however, several key elements are outdated or no longer applicable. The plan stated that MARAD will annually assess and report on its progress and all factors affecting the program, and if necessary, revise the short- and long-term strategy and implementation plan. Since then, however, it has not updated its plan, because, according to program officials, the principles of the plan remain relevant.

MARAD was required, however, to provide regular reports to Congress for several years, the last of which was in March 2011. The reports provided information on the progress made to address the backlog, and other activities accomplished since the last report, but did not provide the strategic short- and long-term direction for the program.[15]

The 2006 strategy does not take into account current market conditions, goals, and external risks. For example, one of MARAD's stated goals was to eliminate the backlog of high priority ships that accumulated in the 1990s. MARAD officials we spoke with acknowledge that the backlog has been addressed and they are now focused on maintaining a steady stream of ships entering the NDRF to be removed from the fleet for disposal. As shown in figure 5, from fiscal years 2005 through 2013, 168 ships were removed from the fleet in preparation for disposal. MARAD has 27 NDRF and 2 Navy ships for disposal at the start of fiscal year 2014 and about three to five obsolete ships will be added to its inventory during fiscal year 2014. Further, MARAD's target is to remove 12 ships for disposal in fiscal year 2014.

The 2006 plan also discusses options for ship disposal that are no longer applicable.

- The plan points to available domestic recycling facilities located along the East and Gulf Coasts; however, as of October 2013, the East Coast facility is no longer considered a qualified facility.
- The plan notes that one of the critical factors that will impact MARAD's goals for ship disposal is making foreign recycling a viable disposal option in 2006 and beyond. Foreign recycling is not an option because the Duncan Hunter National Defense Authorization Act for Fiscal Year 2009 generally prohibited sales to foreign facilities, unless, among other things, no domestic capacity is available.[16]
- The plan describes sales contracts as an option that is of occasional interest and not considered a trend that can be relied upon. MARAD has experienced a trend of sales contracts starting in fiscal year 2011. (See appendix II)

Further, MARAD officials expressed concern about maintaining the supplier base to help ensure adequate competition among contractors as the number of ships available for disposal decreases, a concern that is not addressed in the 2006 plan. Competition is a cornerstone of federal contracting and a critical tool for achieving the best return on the government's investment. Achieving effective competition, in which the government receives more than one offer in response to a solicitation, has been a focus of other agencies, such as the Office of Management and Budget and Department of Defense, because of its noted benefits to the government, including the ability to consider alternative solutions in a reasoned and structured manner.

MARAD officials also expressed concerns about the potential effect on the capacity of three of their qualified facilities that were recently awarded Navy contracts to recycle three large aircraft carriers. According to them, one aircraft carrier is approximately equivalent to eight MARAD ships, and they are concerned that this will reduce the capacity of these facilities which will in turn reduce their availability to compete on future MARAD announcements. *Standards for Internal Control in the Federal Government* call for identifying risks to the agency arising from external and internal factors, and developing a thorough analysis of the possible effect of the risks and an approach for risk management.[17] In addition to updating the current goals and ship disposal options available, MARAD can use the plan to assess future risks and strategies to address internal and external factors, such as supplier base. For example, MARAD has a fiscal year 2011 Memorandum of Agreement with the U.S. Navy to dispose of certain non-combatant ships, a fiscal year 2013 Letter of Agreement with the General Services Administration to disposed of certain federally-owned excess ships, and a Memorandum of Agreement with the U.S. Coast Guard to dispose of certain excess cutters, none of which are included in the 2006 plan. An updated strategic plan that reflects the current external environment and risks could better position MARAD to identify challenges and opportunities confronting the future management of the ship disposal program, including maintaining long- term participation in and competition within the industrial base for ship disposal.

CONCLUSION

Over the last decade, MARAD's ship disposal program has made ample progress disposing of obsolete ships, effectively eliminating the past backlog. MARAD's current process for making source selection decisions is consistent with the FAR's procedures and processes for simplified acquisitions and determining best value. However, MARAD could improve its information sharing with its qualified contractors to be more transparent. Ship recycling contractors have been concerned whether changes in their facility and processes are being approved by MARAD. Contractors were also uncertain about how best value determinations are made. MARAD, on the other hand, believes it is always available to discuss concerns and relies on the contractors to reach out and ask for clarification. Until MARAD improves its communication strategy with contractors, it is not maximizing transparency in its efforts. Moreover, an updated strategic plan or new strategic plan reflecting the current external environment and risks could help MARAD improve its ship disposal program.

RECOMMENDATIONS FOR EXECUTIVE ACTION

To enhance MARAD's transparency in its source selection processes or other aspects of ship disposal and strategic direction for its efforts, we recommend that the Secretary of Transportation direct the MARAD Administrator to take the following two actions:

- Improve MARAD's communication strategy, such as by holding an annual industry day or annual meetings with qualified contractors, to transparently communicate information to qualified contractors and to respond to their questions; and
- Update MARAD's 2006 comprehensive management plan, i.e. strategic plan, or create a new strategic plan to reflect its current goals, external factors affecting the disposal program, and future risks and strategies based on those goals and factors, and periodically update this plan.

AGENCY COMMENTS AND OUR EVALUATION

We provided a draft of this report to DOT for review and comment. DOT did not take a position on our recommendations, but generally agreed with the facts presented. DOT provided technical comments, which we incorporated as appropriate

Marie A. Mak
Acting Director
Acquisition and Sourcing Management Team

APPENDIX I: OBJECTIVES, SCOPE AND METHODOLOGY

The Coast Guard and Maritime Transportation Act of 2012 mandated that GAO conduct an assessment of the source selection procedures and practices used to award the Maritime Administration's (MARAD) ship recycling contracts, including the process, procedures and practices used for qualification of ship disposal facilities, whether MARAD's contract source selection procedures and practices are consistent with the law, and best practices associated with making source selection decisions, as well as any other aspect we deem appropriate to review. This report assesses MARAD's (1) source selection process; (2) communication strategy with ship disposal facility contractors; and (3) long-term ship disposal strategy.

To assess MARAD's source selection process—including facility qualification and best value determination—and its communication strategy, we reviewed federal laws and regulations, guidance, and other documentation associated with the agency's source selection procedures, including simplified acquisition procedures and how best value determinations are addressed in the Federal Acquisition Regulation (FAR). We compared MARAD's procedures to those in the FAR, as applicable. We analyzed its solicitations for fiscal years 2005, 2009, and 2014 that describe the facility qualification process, which represent all the solicitations issued since the agency stopped using Program Research and Development Announcements in 2005. We also reviewed examples of correspondence between MARAD and facilities, specific ship announcements, and required reports and briefings for Congress and other communications with contractor facilities, such as debriefing forms, and recent bid protests related to facility qualification and source selection procedures. We compared MARAD's communication strategy to criteria in *Standards for Internal Control in the Federal Government* related to effective communications.[18] In addition, we reviewed facility qualification documents, e.g. the facilities' submitted general technical plans, MARAD's letters of concern to contractors, and the notification of results from the evaluation of technical plans to determine the length of time it took to become qualified and how monitoring by MARAD was conducted. To identify how MARAD determined best value during the source selection process, we identified and reviewed a random non- representative sample of source selection documents for fiscal year 2010 through 2013, e.g., sales announcements for 18 obsolete ships, facility proposals and MARAD notification memorandums, and contracts, to determine the steps and analysis MARAD uses to select a contractor through a best value process. We selected contracts awarded from fiscal year 2010 to 2013 to include both sales contracts and contracts for services in our sample.

We also interviewed MARAD Ship Disposal Program officials and six contractors representing all seven qualified facilities at the time of our review to determine the processes that are used to select and award contracts. We also interviewed one contractor with one facility going through the qualification process at the time of our review. Through interviews with these officials, we determined (1) how the qualification process was applied for facilities to become technically acceptable and qualified; (2) how MARAD monitors a facility's qualification status and any changes to the facility; (3) the process for source selection, including how best value determinations are made for specific ship announcements for disposal; (4) how MARAD communicates its source selection process with contractors; and (5) challenges identified by facilities or MARAD relating to the award of ship disposal contracts.

MARAD's pool of qualified ship disposal facilities at the time of GAO review included:[19]

1. All Star Metals, LLC
2. Mare Island Shipyard, LLC (formerly Allied Defense Recycling, LLC)
3. Bay Bridge Texas, LLC (undergoing qualification)
4. BB Metals Enterprises, Inc.
5. Esco Marine, Inc.
6. International Shipbreaking Ltd.
7. Marine Metals, Inc.
8. Southern Recycling, LLC

To assess MARAD's long-term strategy for its ship disposal program, we reviewed its current strategy—the fiscal year 2006 Comprehensive Management Plan for the Disposal of MARAD Non-Retention Vessels. In reviewing MARAD's most recent strategy, we identified its decision making framework and its schedule and milestones for ship disposal, which included planning up through fiscal year 2007. We reviewed *Standards for Internal Control in the Federal Government* related to strategic planning, and assessed MARAD's actions against these standards. We also reviewed MARAD's agreements with other agencies to dispose of their obsolete ships, e.g., its memorandums of agreement with the U.S. Navy, General Services Administration, and U.S. Coast Guard to dispose of certain ships. We also obtained data from MARAD about the number of obsolete ships on hand and removed for disposal through MARAD's Ship Disposal Program for fiscal years' 2005 through 2013. We spoke with MARAD officials and facility contractors to identify concerns related to MARAD's strategy, including supplier base concerns for the ship recycling industry, and how it communicates its future ship disposal plans with qualified facilities.

We conducted this performance audit from May 2013 to February 2014 in accordance with generally accepted government auditing standards. Those standards require that we plan and perform the audit to obtain sufficient, appropriate evidence to provide a reasonable basis for our findings and conclusions based on our audit objectives. We believe that the evidence obtained provides a reasonable basis for our findings and conclusions based on our audit objectives.

Appendix II: Maritime Administration's Ship Disposal Options

The Maritime Administration's (MARAD) available ship disposal options include: domestic recycling, artificial reefing, deep-water sinking (called SINKEX), donations, and sale for reuse.

- **Domestic recycling**—MARAD sells the ships to qualified contractors or purchases recycling services from these contractors.
- **Artificial reefing**—MARAD accepts applications from coastal states, U.S. territories and possessions, and foreign governments for use of obsolete NDRF ships as offshore reefs for the conservation of marine life.
- **SINKEX**—Navy and MARAD jointly operate ship disposal projects through the Navy's sink at-sea live-fire training exercises (SINKEX Program).
- **Ship donations**—MARAD may convey obsolete NDRF ships to a non-profit organization, state, commonwealth, or U.S. possession for ship reutilization including historic restoration of ships as memorials and museums, and the operational restoration of ships to support non-profit humanitarian missions.
- **Sale for reuse**—MARAD sells ships to another buyer for reuse.

Type of disposal	2001	2002	2003	2004	2005	2006	2007	2008	2009	2010	2011	2012	2013	Totals
	Fiscal year													
Recycling (fee for service)	3	4	2	13	17	19	12	9	9	10	13			**111**
Recycling (sales)	1	2		2	1	5	3	15	5		8	16	17	**75**
Reefing	1						2			1				**4**
SINKEX						1				1				**2**
Donation	1							1						**2**
Sale for reuse							3							**3**
Totals	**6**	**6**	**2**	**15**	**18**	**25**	**20**	**25**	**14**	**12**	**21**	**16**	**17**	**197**

Source: GAO analysis of MARAD information.

Figure 6. Number of Ships the Maritime Administration (MARAD) Removed from the Fleet for Disposal, by Type, Fiscal Years 2001 through 2013.

Appendix III: Maritime Administration's Process for Qualifying Ship Disposal Facilities

Table 1 presents additional information regarding Figure 1.

Table 1. Maritime Administration' s (MARAD) Qualification Process of Ship Disposal Facilities

Step in qualification process	Information presented in step (interactive format)
circle: **Are proposals complete and feasible?**	MARAD conducts an initial evaluation to determine if the proposal is complete, meaning it complies with the solicitation requirements and contains enough information to undergo evaluation. According to MARAD officials, most initial proposals are rarely complete and require iterative back and forth communication with the contractor to obtain sufficient information. Once complete, MARAD determines whether the proposal is feasible for further evaluation. Once proposal is considered feasible, MARAD will schedule a pre-qualification site visit.
box 1: **Proposal deficiencies communicated in writing to contractor**	MARAD requires facilities to demonstrate their ship disposal capability based on several criteria. Evaluation criteria of proposals include: feasibility and likelihood of success of engineering/technical/ management approach for disposal of obsolete ships and associated risks; productivity of proposed approach, including schedule/cost; type and sufficiency of proposed methods, processes, and procedures; depth, relevance, and currency of requisite experience. According to MARAD officials, qualification criteria have not changed since 2005.
box 2: **MARAD conducts site visit and pre- qualification survey with facility**	MARAD conducts a pre-qualification survey during the facility site visit to verify the proposal's information—such as the facility infrastructure, capacity, and the hazardous material and recycling processes capability—and to evaluate the contractor's stated capabilities. As part of the pre-qualification

Step in qualification process	Information presented in step (interactive format)
	survey, MARAD assesses the following areas relevant to the submitted proposal: (1) the feasibility of the engineering, technical, and management approach to conduct ship recycling activities, including the risks associated with the proposed approach; (2) the productivity of the proposed ship recycling approach, including the reasonableness of the performance schedule; (3) the type and sufficiency of proposed methods, processes and procedures for environmental abatement of hazardous materials, worker safety and health considerations proposed, and compliance with applicable laws; and (4) the depth, relevance and currency of requisite experience with the complete disposal of obsolete ships. As part of the assessment of experience, MARAD meets with the facility owners, management team, and major subcontractors.
box 3: **MARAD completes proposal evaluation and reaches a qualification decision**	A qualification determination is reached by MARAD in one of three possible categories: • Technically unacceptable: no further consideration, but contractor can reapply with a new proposal; • Technically unacceptable: but susceptible to being made technically acceptable (TUSTA): MARAD informs contractor of deficiencies; or • Technically acceptable: contractor is included in the pool of qualified facilities.

Source: GAO analysis of MARAD information.

End Notes

[1] Established in 1946, the National Defense Reserve Fleet serves as a reserve that could be activated to meet shipping requirements during national emergencies. The fleet contains some older ships that have been designated for disposal (referred to as non- retention ships). In this report, we refer to non-retention ships as obsolete ships.

[2] Floyd D. Spence National Defense Authorization Act for Fiscal Year 2001, Pub. L. No. 106-398, § 3502 (2000). For the purposes of this report, we generally refer to sales contracts and contracts for services as recycling contracts.

[3] An agency can obtain best value by using certain source selection approaches. MARAD specifically uses a tradeoff process which permits tradeoffs among cost or price and non-cost factors and allows the Government to accept other than the best priced offer. FAR § 15.101. See FAR § 2.101 for the definition of best value.

[4] For the purposes of this report, when we use the term "best price" we mean an award of a sales contract to the highest priced offer, or an award of a contract for services to the lowest priced offer.

[5] Pub. L. No. 112-213, § 415.

[6] Two of the contractors we interviewed no longer have a qualified facility.

[7] GAO, *Standards for Internal Control in the Federal Government*, GAO/AIMD-00-21.3.1 (Washington, D.C. Nov. 1999).

[8] Specifically, MARAD uses the Commercial Item Test Program, FAR Subpart 13.5. This generally authorizes the use of simplified procedures for the acquisition of services in amounts under $6.5M, including options, if the contracting officer reasonably expects that offers will include only commercial items.

[9] Prior to 2005, MARAD used Program Research and Development Announcements (PDRA) to award ship disposal contracts. In 2005, we reported that MARAD's use of PDRAs was inappropriate because PDRAs are intended to be used to acquire research or development services. See GAO, *Maritime Administration: Improved Program Management Needed to Address Timely Disposal of Obsolete Ships,* GAO-05-264, (Washington, D.C.: Mar. 7, 2005).

[10] This facility is no longer qualified. According to MARAD, the contractor decided not to respond to the fiscal year 2014 solicitation to remain a qualified facility and to instead focus on other business opportunities.

[11] FAR §13.106-1(a)(2) allows the agency not to specify the relative order of importance of the evaluation factors. Under negotiated contracting processes, agencies are required to provide a relative order of importance of the evaluation criteria.

[12] To determine that MARAD's process is consistent with the FAR, we considered the general processes that we observed in our review of selected contract awards and discussed these with MARAD. We did not consider how this process is applied in the context of a specific solicitation or contract award. Consequently, this determination does not represent a decision by GAO's bid protest function.

[13] GAO/AIMD-00-21.3.1.

[14] GAO-05-264.

[15] In 2011, Congress amended the reporting requirement to simply require MARAD to provide briefings upon request. National Defense Authorization Act for Fiscal Year 2012, Pub. L. No. 112-81, § 3504 (2011).

[16] Pub. L. No. 110-417, § 3502 (2008).

[17] GAO/AIMD-00-21.3.1.

[18] GAO, *Standards for Internal Control in the Federal Government*, GAO/AIMD-00-21.3.1 (Washington, D.C. Nov. 1999).

[19] Mare Island Shipyard in Vallejo, Calif. and BB Metals Enterprise, Inc. in Baltimore, Md. are no longer qualified. Both facilities chose not to submit new general technical proposals as required under the new ship disposal solicitation. International Shipbreaking Ltd. and Southern Recycling, LLC are owned by the same contractor. Also, according to MARAD officials, since the time of our review, one additional facility owned by this contractor became qualified. To summarize, as of December 2013, there were six qualified facilities owned by four contractors.

In: Federal Disposal Responsibilities
Editor: Tara C. Waller
ISBN: 978-1-63321-217-6

Chapter 3

CIVILIAN NUCLEAR WASTE DISPOSAL*

Mark Holt

SUMMARY

Management of civilian radioactive waste has posed difficult issues for Congress since the beginning of the nuclear power industry in the 1950s. Federal policy is based on the premise that nuclear waste can be disposed of safely, but proposed storage and disposal facilities have frequently been challenged on safety, health, and environmental grounds. Although civilian radioactive waste encompasses a wide range of materials, most of the current debate focuses on highly radioactive spent fuel from nuclear power plants. The United States currently has no disposal facility for spent nuclear fuel.

The Nuclear Waste Policy Act of 1982 (NWPA) calls for disposal of spent nuclear fuel in a deep geologic repository. NWPA established the Office of Civilian Radioactive Waste Management (OCRWM) in the Department of Energy (DOE) to develop such a repository, which would be licensed by the Nuclear Regulatory Commission (NRC). Amendments to NWPA in 1987 restricted DOE's repository site studies to Yucca Mountain in Nevada. DOE

* This is an edited, reformatted and augmented version of a Congressional Research Service publication RL33461, prepared for Members and Committees of Congress dated December 2, 2013.

submitted a license application for the proposed Yucca Mountain repository to NRC on June 3, 2008. The state of Nevada strongly opposes the Yucca Mountain project, citing excessive water infiltration, earthquakes, volcanoes, human intrusion, and other technical flaws.

The Obama Administration "has determined that developing the Yucca Mountain repository is not a workable option and the Nation needs a different solution for nuclear waste disposal," according to the DOE FY2011 budget justification. As a result, no funding for Yucca Mountain, OCRWM, or NRC licensing was requested or provided for FY2011 or subsequent years. DOE filed a motion with NRC to withdraw the Yucca Mountain license application on March 3, 2010. An NRC licensing board denied DOE's withdrawal motion on June 29, 2010, a decision sustained by the NRC commissioners on a tie vote September 9, 2011. Despite that decision, NRC halted further consideration of the license application because of "budgetary limitations," but a federal appeals court on August 13, 2013, ordered NRC to continue the licensing process with previously appropriated funds.

After halting the Yucca Mountain project, the Administration established the Blue Ribbon Commission on America's Nuclear Future to develop an alternative nuclear waste policy. The commission issued its final report on January 26, 2012, recommending that a new, "single-purpose organization" be given the authority and resources to promptly begin developing one or more nuclear waste repositories and consolidated storage facilities. The commission recommended a "consent based" process for siting nuclear waste storage and disposal facilities and that long-term research, development, and demonstration be conducted on technologies that could provide waste disposal benefits.

After OCRWM was dismantled, responsibility for implementing the Administration's nuclear waste policy was given to DOE's Office of Nuclear Energy (NE). In January 2013, NE issued a nuclear waste strategy based on the Blue Ribbon Commission recommendations. The strategy calls for a pilot interim storage facility for spent fuel from closed nuclear reactors to open by 2021 and a larger storage facility, possibly at the same site, to open by 2025. A site for a permanent underground waste repository would be selected by 2026, and the repository would open by 2048. DOE requested $60 million for FY2014 to carry out the new waste strategy, an amount approved by the Senate Appropriations Committee. However, the House denied funding for the new strategy and provided $25 million for Yucca Mountain instead.

Most Recent Developments

The Obama Administration's nuclear waste policy calls for termination of the Yucca Mountain repository project and the development of alternative approaches to waste management. Under the Nuclear Waste Policy Act (NWPA), the Yucca Mountain site in Nevada has been the only location under consideration by the Department of Energy (DOE) for construction of a national high-level radioactive waste repository. DOE had submitted a license application for the Yucca Mountain repository to the Nuclear Regulatory Commission (NRC) on June 3, 2008, but the Obama Administration has halted funding for the project and moved to withdraw the application.

An NRC licensing board denied DOE's withdrawal motion on June 29, 2010, a decision sustained by the NRC commissioners on a tie vote September 9, 2011. Despite that decision, NRC halted further consideration of the license application because of "budgetary limitations," but a federal appeals court on August 13, 2013, ordered NRC to continue the licensing process with previously appropriated funds. NRC responded November 18, 2013, by directing the agency's staff to complete the Yucca Mountain safety evaluation report, a key document that would provide the staff's conclusions about whether the proposed repository could be licensed. NRC also requested DOE to prepare a supplemental environmental impact statement that NRC needs to complete its environmental review of the Yucca Mountain application.

After taking action to terminate the Yucca Mountain project, the Administration established the Blue Ribbon Commission on America's Nuclear Future (BRC) to develop an alternative nuclear waste policy. The commission issued its final report on January 26, 2012, recommending that a new, "single-purpose organization" be given the authority and resources to promptly begin developing one or more nuclear waste repositories and consolidated storage facilities. The recommendations called for a "consent based" process in which the roles of various levels of government in siting and regulating nuclear waste facilities would be established through negotiations. The commission also recommended that long-term research, development, and demonstration be conducted on technologies that could provide waste disposal benefits.

In response to the BRC report, and to provide an outline for a new nuclear waste program, DOE issued a *Strategy for the Management and Disposal of Used Nuclear Fuel and High-Level Waste* in January 2013. The DOE strategy calls for a new nuclear waste management entity to develop consent-based storage and disposal sites, similar to the BRC recommendation. Under the DOE strategy, a pilot interim spent fuel storage facility would be opened by 2021 and a larger-scale storage facility, which could be an expansion of the pilot facility, by 2025. A geologic disposal facility would open by 2048—50 years after the initially planned opening date for the Yucca Mountain repository.[1]

President Obama's FY2014 federal budget, submitted to Congress April 11, 2013, included no funding for Yucca Mountain but requested $60 million "to support the program of work" outlined by DOE's January 2013 nuclear waste strategy. The House provided none of the requested nuclear waste strategy funding in the FY2014 Energy and Water Development appropriations bill (H.R. 2609, H.Rept. 113-135), passed July 10, 2013. Instead, the House included $25 million in the bill for work related toYucca Mountain. The Senate Appropriations Committee endorsed the Administration's nuclear waste funding request in its version of the Energy and Water Bill (S. 1245, S.Rept. 113-47), approved June 27, 2013. FY2014 nuclear waste funding is currently provided through January 15, 2014, by a continuing resolution (P.L. 113-46), at about the spending rate requested by the Administration and with no new funds for Yucca Mountain.

Citing uncertainty about the future of the nuclear waste program, the U.S. Court of Appeals for the District of Columbia Circuit ordered DOE on November 19, 2013, to stop collecting fees on nuclear power that are supposed to pay for waste disposal. The fees, authorized by NWPA, are paid by nuclear power generators at the rate of a tenth of a cent per kilowatt-hour and total about $800 million per year. NWPA requires the Secretary of Energy to adjust the fees as necessary to cover the waste program's anticipated costs, but the Court ruled that DOE's current waste plans are too vague to allow a reasonable estimate to be calculated. The Court noted that DOE's most recent cost estimate for the program had an uncertainty range of nearly $7 trillion, a range "so large as to be absolutely useless" for determining the waste fee.[2]

On June 8, 2012, the Court of Appeals overturned NRC's Waste Confidence Decision, which determined that spent nuclear fuel could be safely stored at nuclear reactor sites for at least 60 years after they have shut down and that permanent disposal would be available "when necessary." The court ruled that NRC should have conducted an environmental review under the National Environmental Policy Act before issuing the Waste Confidence Decision in December 2010.[3] Under previous court rulings, NRC must determine that waste from proposed nuclear plants can be safely managed before licensing them to operate. As a result, NRC has halted licensing of new facilities that rely on the Waste Confidence Decision, and it plans to complete an environmental impact statement and issue a new Waste Confidence rule by September 2014.[4] The NRC commissioners approved a proposed Waste Confidence rule on August 5, 2013.[5]

The March 11, 2011, disaster at Japan's Fukushima Daiichi nuclear power plant increased concern about spent fuel stored in pools of water at nuclear plant sites. The loss of power at the site, caused by a huge earthquake and tsunami, disabled cooling systems at the plant's spent fuel pools. Water in the pools was initially feared to have boiled or leaked and dropped below the level of the stored spent fuel, but later analysis indicated that the spent fuel did not overheat. Nevertheless, the incident has prompted numerous recommendations for safety improvements at spent fuel pools. NRC approved an order March 9, 2012, requiring U.S. reactors to install improved water-level monitoring equipment at their spent fuel pools.[6]

INTRODUCTION

Nuclear waste has sometimes been called the Achilles'heel of the nuclear power industry. Much of the controversy over nuclear power centers on the lack of a disposal system for the highly radioactive spent fuel that must be regularly removed from operating reactors. Low-level radioactive waste generated by nuclear power plants, industry, hospitals, and other activities is also a long-standing issue.

Spent Nuclear Fuel Program

The Nuclear Waste Policy Act of 1982 (NWPA, P.L. 97-425), as amended in 1987, required the Department of Energy (DOE) to focus on Yucca Mountain, Nevada, as the site of a deep underground repository for spent nuclear fuel and other highly radioactive waste. The state of Nevada has strongly opposed DOE's efforts on the grounds that the site is unsafe, pointing to potential volcanic activity, earthquakes, water infiltration, underground flooding, nuclear chain reactions, and fossil fuel and mineral deposits that might encourage future human intrusion.

Under the George W. Bush Administration, DOE determined that Yucca Mountain was suitable for a repository and that licensing of the site by the

Nuclear Regulatory Commission (NRC) should proceed, as specified by NWPA. DOE submitted a license application for the repository to NRC on June 3, 2008, and projected that the repository could begin receiving waste in 2020, about 22 years later than the 1998 goal established by NWPA.[7]

However, the Obama Administration made a policy decision that the Yucca Mountain repository should not be opened, largely because of Nevada's continuing opposition, although it requested FY2010 funding to continue the NRC licensing process. But the Administration's FY2011 budget request reversed the previous year's plan to continue licensing the repository and called for a complete halt in funding and elimination of the Office of Civilian Radioactive Waste Management (OCRWM), which had run the program. In line with the request, the FY2011 Continuing Appropriations Act (P.L. 112-10) provided no DOE funding for the program. DOE shut down the Yucca Mountain project at the end of FY2010 and transferred OCRWM's remaining functions to the Office of Nuclear Energy.

DOE filed a motion to withdraw the Yucca Mountain license application on March 3, 2010, "with prejudice," meaning the application could not be resubmitted to NRC in the future.[8] DOE's motion to withdraw the license application, filed with NRC's Atomic Safety and Licensing Board (ASLB), received strong support from the state of Nevada but drew opposition from states with defense-related and civilian radioactive waste that had been expected to go to Yucca Mountain. State utility regulators also filed a motion to intervene on March 15, 2010, contending that "dismissal of the Yucca Mountain application will significantly undermine the government's ability to fulfill its outstanding obligation to take possession and dispose of the nation's spent nuclear fuel and high level nuclear waste."[9]

The ASLB denied DOE's license withdrawal motion June 29, 2010, ruling that the NWPA prohibits DOE from withdrawing the license application until NRC determines whether the repository is acceptable.[10] The NRC commissioners sustained the ASLB decision on a tie vote September 9, 2011. However, NRC halted further consideration of the license application because of "budgetary limitations."[11] Lawsuits to overturn the Yucca Mountain license withdrawal on statutory grounds were filed with the U.S. Court of Appeals for the District of Columbia Circuit, which ruled on August 13, 2013, that NRC must continue work on the Yucca Mountain license application as long as funding is available. The Court determined that NRC has at least $11.1 million in previously appropriated funds for that purpose.[12]

NRC responded November 18, 2013, by directing the agency's staff to complete the Yucca Mountain safety evaluation report (SER), a key document that would provide the staff's conclusions about whether the proposed repository could be licensed. NRC also requested DOE to prepare a supplemental environmental impact statement that NRC needs to complete its environmental review of the Yucca Mountain application. Documents previously used in the licensing process will be loaded into NRC's database and made public if they are used for the SER. Additional documents may also be made public if funds are available. However, NRC does not plan to resume the adjudication of Yucca Mountain licensing issues.[13]

To develop alternative waste disposal strategies, the Administration established the Blue Ribbon Commission on America's Nuclear Future (BRC). The BRC issued its final report on January 26, 2012, recommending that a new, "single-purpose organization" be given the authority and resources to promptly begin developing one or more nuclear waste repositories and consolidated storage facilities. The new organization would use a "consent based" process to select waste facility sites.[14] The BRC had commissioned a series of reports on various aspects of nuclear waste policy to assist in its deliberations.[15] In response to the BRC report, and to provide an outline for a new nuclear waste program, DOE issued its *Strategy for the Management and Disposal of Used Nuclear Fuel and High-Level Waste* in January 2013. The DOE strategy calls for a new nuclear waste management entity to develop consent-based storage and disposal sites, similar to the BRC recommendation. Under the DOE strategy, a pilot interim spent fuel storage facility would be opened by 2021 and a larger-scale storage facility, which could be an expansion of the pilot facility, by 2025. A geologic disposal facility would open by 2048—50 years after the initially planned opening date for the Yucca Mountain repository.[16]

The safety of geologic disposal of spent nuclear fuel and high-level waste (HLW), as planned in the United States, depends largely on the characteristics of the rock formations from which a repository would be excavated. Because many geologic formations are believed to have remained undisturbed for millions of years, it appeared technically feasible to isolate radioactive materials from the environment until they decayed to safe levels. "There is strong worldwide consensus that the best, safest long-term option for dealing with HLW is geologic isolation," according to the National Research Council.[17]

But, as the Yucca Mountain controversy indicates, scientific confidence about the concept of deep geologic disposal has turned out to be difficult to apply to specific sites. Every high-level waste site that has been proposed by DOE and its predecessor agencies has faced allegations or discovery of unacceptable flaws, such as water intrusion or earthquake vulnerability, that could release unacceptable levels of radioactivity into the environment. Much of the problem results from the inherent uncertainty involved in predicting waste site performance for the 1 million years that nuclear waste is to be isolated under current regulations. Widespread public controversy has also arisen over potential waste transportation routes to the sites under consideration.

President Obama's budgets for FY2014 and previous years have included long-term research on a wide variety of technologies that could reduce the volume and toxicity of nuclear waste. The Bush Administration had proposed to demonstrate large-scale facilities to reprocess and recycle spent nuclear fuel by separating long-lived elements, such as plutonium, that could be made into new fuel and "transmuted" into shorter-lived radioactive isotopes. Spent fuel reprocessing, however, has long been controversial because of the potential weapons use of separated plutonium and cost concerns. The Obama Administration has refocused DOE's nuclear waste research toward fundamental science and away from the near-term design and development of reprocessing facilities.

President Bush had recommended the Yucca Mountain site to Congress on February 15, 2002, and Nevada Governor Guinn submitted a notice of disapproval, or "state veto," April 8, 2002, as allowed by NWPA. The state veto would have blocked further repository development at Yucca Mountain if a resolution approving the site had not been passed by Congress and signed into law within 90 days of continuous session. An approval resolution was signed by President Bush July 23, 2002 (P.L. 107-200).[18]

Other Programs

Other types of civilian radioactive waste have also generated public controversy, particularly low-level waste, which is produced by nuclear power plants, medical institutions, industrial operations, and research activities. Civilian low-level waste currently is disposed of in large trenches at sites in the states of South Carolina, Texas, and Washington. However, the Washington facility does not accept waste from outside its region, and the South Carolina site is available only to the three members of the Atlantic disposal compact (Connecticut, New Jersey, and South Carolina) as of June 30, 2008. The lowest-concentration class of low-level radioactive waste (class

A) is accepted by a Utah commercial disposal facility from anywhere in the United States.

Threats by states to close their disposal facilities led to congressional authorization of regional compacts for low-level waste disposal in 1985. The first, and so far only, new disposal site under the regional compact system opened on November 10, 2011, near Andrews, TX.[19] The Texas Legislature approved legislation in May 2011 to allow up to 30% of the facility's capacity to be used by states outside the Texas Compact, which consists of Texas and Vermont.[20]

NUCLEAR WASTE LITIGATION

NWPA Section 302 authorized DOE to enter into contracts with U.S. generators of spent nuclear fuel and other highly radioactive waste; under the contracts, DOE was to dispose of the waste in return for a fee on nuclear power generation. The act prohibited nuclear reactors from being licensed to operate without a nuclear waste disposal contract with DOE, and all reactor operators subsequently signed them. As required by NWPA, the "standard contract" specified that DOE would begin disposing of nuclear waste no later than January 31, 1998.[21]

After DOE missed the contractual deadline, nuclear utilities began filing lawsuits to recover their additional storage costs—costs they would not have incurred had DOE begun accepting waste in 1998 as scheduled. DOE reached its first settlement with a nuclear utility, PECO Energy Company (now part of Exelon), on July 19, 2000. The agreement allowed PECO to keep up to $80 million in nuclear waste fee revenues during the subsequent 10 years. However, other utilities sued DOE to block the settlement, contending that nuclear waste fees may be used only for the DOE waste program and not as compensation for missing the disposal deadline. The U.S. Court of Appeals for the 11th Circuit agreed, ruling September 24, 2002, that any compensation would have to come from general revenues or other sources than the waste fund.

Through the end of September 2012, nuclear waste payments from the Judgment Fund totaled $2.6 billion from settlements as well as court awards, according to one of the primary attorneys involved in the litigation. Of the 74 lawsuits filed for waste delays, 31 had been settled, 16 resulted in court awards, and 27 were pending.[22] The payments resulting from the court judgments and settlements are made from the U.S. Treasury's Judgment Fund, a permanent account that is used to cover damage claims against the U.S. government. Under the settlements, utilities submit annual reimbursement claims to DOE for any delay-related nuclear waste storage costs they incurred during that year. Any disagreements over reimbursable claims between DOE and a utility would go to arbitration.

Utilities that have not settled with the Department of Justice have continued seeking damage compensation through the U.S. Court of Federal Claims. Unlike the settlements, which cover all past and future damages resulting from DOE's nuclear waste delays, awards by the Court of Claims can cover only damages that have already been incurred; therefore, utilities must continue filing claims as they accrue additional delay-related costs.

Future Liability Estimates

DOE estimates that its potential liabilities for waste program delays would total as much as $23 billion during the next 50 years (in current dollars) if the department were able to begin taking spent nuclear fuel from plant sites by 2020, which had been the most recent goal under the Bush Administration.[23] DOE's methodology for estimating its nuclear waste liability is shown in **Figure 1** (although the liability estimate has more than doubled since this graph was issued in 2008). The yellow line shows DOE's estimate of how much spent fuel would have been removed from nuclear plant sites had shipments begun on the NWPA deadline of January 1998. The rate of waste acceptance under that scenario is 900 metric tons per year from 1998 through 2015 and 2,100 tons/year thereafter. That assumed acceptance rate was negotiated by DOE as part of the settlements discussed above. The annual costs reimbursed by DOE under the settlements cover utilities'expenses for storing waste that would have already been taken away under the assumed acceptance rate (the yellow line).

The green and red lines in **Figure 1** show DOE's planned waste acceptance rate if waste shipments were to begin by 2017 or 2020. Under those scenarios, DOE would take away 400 metric tons the first year, 600 the second year, 1,200 the third year, 2,000 the fourth year, and 3,000 per year thereafter. This is the rate assumed by DOE's Total System Life Cycle Cost Report.[24] At that higher acceptance rate, DOE would be able to eventually catch up with the amount of waste that it was assumed to take under the settlements (the yellow line). If waste acceptance began by 2017 (the green line), the backlog would be eliminated by 2046, and if acceptance began by 2020 (the red line) the backlog would be gone by 2056. Under the settlements, therefore, there would be no further annual damage payments after those years, if DOE were able to achieve the 2017 or 2020 acceptance scenario.

DOE bases its estimate of the total damage payments that would be paid through 2046 or 2056 on the amounts paid to date under the settlement claims. As noted above, estimates of future payments have already risen substantially since the 2008 estimates shown in **Figure 1**. If damage awards by the Court of Claims (currently involving about two-thirds of U.S. reactors) exceed the rates paid under the settlements, then future payments could further exceed those estimates.

Further delays in the start of waste acceptance would delay the point at which DOE would catch up to the cumulative waste shipments assumed under the settlement scenario (yellow line) and would no longer have to make annual damage payments. DOE estimates that each year's delay in the startup date would increase the total eventual damage payments by as much as $500 million.[25]

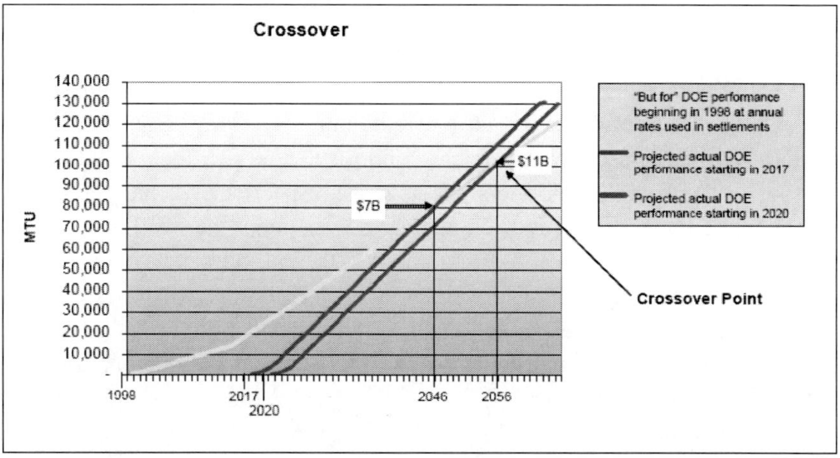

Source: Christopher A. Kouts, Principal Deputy Director, Office of Civilian Radioactive Waste Management, U.S. Department of Energy, "Yucca Mountain Program Status Update," July 22, 2008, p. 18.

Note: MTU stands for metric tons of uranium in the initial fuel. MTU in the graph are cumulative.

Figure 1. DOE Estimate of Future Liabilities for Nuclear Waste Delays.

DOE filed a license application with the Nuclear Regulatory Commission (NRC) for the proposed Yucca Mountain repository in June 2008, and has estimated that annual program spending would have to increase to nearly $2 billion (from around $300 million in FY2009) to allow waste shipments to begin by 2020 if the license were approved.[26] However, funding for the Yucca Mountain project was halted in FY2011, as noted above. As a result, it appears unlikely that spent nuclear fuel shipments to Yucca Mountain could begin by 2020, even if full funding for the project were to be restored in the future. Waste acceptance by 2020 might be possible if Congress were to authorize one or more temporary storage sites within the next few years, although previous efforts to develop such facilities have been blocked by state and local opposition. As noted above, DOE's January 2013 waste strategy calls for a pilot storage site to open by 2022 and a full-scale disposal facility by 2025.

Delays in the federal waste disposal program could also lead to future environmental enforcement action over DOE's own high-level waste and spent fuel, mostly resulting from defense and research activities. Some of the DOE-owned waste is currently being stored in non-compliance with state and federal environmental laws, making DOE potentially subject to fines and penalties if the waste is not removed according to previously negotiated compliance schedules.

The National Association of Regulatory Utility Commissioners (NARUC), representing state utility regulators, and the Nuclear Energy Institute (NEI), representing the nuclear industry, filed petitions with the U.S. Court of Appeals on April 2 and April 5, 2010, respectively, to halt the federal government's collection of fees on nuclear power under the NWPA contracts. The suits argue that the fees, totaling about $800 million per year, should not be collected while the federal government's nuclear waste disposal program has been halted.[27] DOE responded that the federal government still intends to dispose of the nation's nuclear waste and that the fees must continue to be collected to cover future disposal costs.[28] Energy Secretary Steven Chu issued a formal determination on November 1, 2010, that there was "no reasonable basis at this time" to conclude that excess funds were being collected for future nuclear waste disposal activities.[29]

The U.S. Court of Appeals for the District of Columbia Circuit ruled June 1, 2012, that Secretary Chu's determination that the nuclear waste fee should continue unchanged was not "a valid evaluation" and ordered him to conduct a more thorough study of the fee within six months. The court noted that the Secretary's finding relied primarily on costs that had been projected for the Yucca Mountain site, which the Administration had terminated as "unworkable." The court concluded that the Secretary must evaluate the likely costs of reasonable alternatives and the timing of those costs, all of which would affect the level of nuclear waste fees required.[30]

DOE responded with a new fee adequacy assessment in January 2013 that evaluated the total costs of a variety of waste management scenarios. The costs of some scenarios exceeded projected revenues from the existing waste fee by as much as $2 trillion, but other scenarios resulted in a surplus of up to $5 trillion. Because of the widely varying results, DOE concluded that there was no clear evidence that the fee should be immediately raised or lowered.[31]

After NEI and NARUC asked for a review of DOE's latest fee adequacy assessment, the Circuit Court ordered DOE on November 19, 2013, to stop collecting the nuclear waste fees altogether. The Court ruled that DOE's current waste plans were too vague to allow a reasonable estimate to be calculated. The Court noted that DOE's $7 trillion uncertainty range for the program's cost was "so large as to be absolutely useless" for determining the waste fee.[32]

License Withdrawal

DOE's motion to withdraw the Yucca Mountain license application "with prejudice," meaning that it could not be resubmitted in the future, was filed with NRC's Atomic Safety and Licensing Board (ASLB) on March 3, 2010. DOE's motion argued that the licensing process should be terminated because "the Secretary of Energy has decided that a geologic repository at Yucca Mountain is not a workable option" for long-term nuclear waste disposal. Subsequent DOE statements have reiterated that the license withdrawal motion was not based on scientific or technical findings. Instead, the policy change was prompted by the perceived difficulty in overcoming continued opposition from the state of Nevada and a desire to find a waste solution with greater public acceptance, according to DOE.[33] DOE contended that the license application should be withdrawn "with prejudice" because of the need to "provide finality in ending the Yucca Mountain project."[34]

The state of Nevada strongly endorsed DOE's motion to withdraw the license application with prejudice[35] and has moved to intervene in a court challenge to the license withdrawal.[36] Nevada has long contended that the geology of the site is unsuitable for long-term nuclear waste disposal.

However, DOE's withdrawal motion has drawn opposition from states and localities with defense-related and civilian nuclear waste that had been expected to go to Yucca Mountain. The state of South Carolina, which has large amounts of high-level radioactive waste at DOE's Savannah River Site, and the state of Washington, which hosts extensive nuclear waste storage facilities at DOE's Hanford Site, filed motions to intervene in the Yucca Mountain licensing proceeding to oppose the license application withdrawal.

NARUC also filed a motion to intervene in the Yucca Mountain licensing proceedings, contending that "dismissal of the Yucca Mountain application will significantly undermine the government's ability to fulfill its outstanding obligation to take possession and dispose of the nation's spent nuclear fuel and high level nuclear waste." NARUC's motion also contends that $17 billion collected from utility ratepayers for the nuclear waste program will be wasted if the Yucca Mountain license application is withdrawn.[37] Also seeking to intervene were Aiken County, SC, and the Prairie Island Indian Community in Minnesota.

The ASLB denied DOE's license withdrawal motion June 29, 2010, ruling that NWPA prohibits DOE from withdrawing the license application until NRC determines whether the repository is acceptable. According to the board, "Surely Congress did not contemplate that, by withdrawing the Application, DOE might unilaterally terminate the Yucca Mountain review process in favor of DOE's independent policy determination that 'alternatives will better serve the public interest.'"[38]

In appealing the ASLB decision to the NRC commissioners, DOE argued in a July 9, 2010, brief that the Secretary of Energy has broad authority under the Atomic Energy Act and Department of Energy Organization Act "to make policy decisions regarding disposal of nuclear waste and spent nuclear fuel." DOE contended that such authority includes "the authority to discontinue the Yucca Mountain project" and that NRC rules provide "that applicants in NRC licensing proceedings may withdraw their applications."[39] After more than a year of deliberation, the NRC commissioners sustained the licensing board's decision on a tie vote September 9, 2011. However, NRC halted further consideration of the license application because of "budgetary limitations."[40]

South Carolina and Aiken County filed challenges to the Yucca Mountain license withdrawal in the U.S. Court of Appeals for the District of Columbia Circuit, contending that NWPA requires the licensing process to proceed. After NRC rejected the license withdrawal motion, the plaintiffs in that case, including Nye County, NV, where Yucca Mountain is located, petitioned the court to order NRC to continue the licensing proceedings, as noted above.[41] The court held oral arguments in the case May 2, 2012. NRC filed an update with the court on January 4, 2013, contending that the reelection of President

Obama and the exclusion of additional Yucca Mountain funding from the most recent FY2013 Continuing Resolution reinforced NRC's suspension of further action on the repository license.[42]

The Court of Appeals ruled on August 13, 2013, that NRC must continue work on the Yucca Mountain license application as long as funding is available. The Court determined that NRC has at least $11.1 million in previously appropriated funds for that purpose.[43]

Waste Confidence Decision

Before issuing licenses to nuclear reactors and waste storage facilities, NRC is required by a 1979 court decision to determine that waste from those facilities can be safely disposed of.[44] To meet that requirement, NRC issued a Waste Confidence Decision in 1984 that found that nuclear waste could be safely stored at reactor sites for at least 30 years after plant closure and that a permanent repository would be available by 2007-2009.[45] At that time, DOE officially planned to meet the NWPA repository deadline of 1998.

After DOE's schedule for opening a nuclear waste repository began to slip, NRC updated the Waste Confidence Decision in 1990 to find that a repository would be available by the first quarter of the next century.[46] When the Yucca Mountain repository was delayed further and then terminated by the Obama Administration, NRC issued another Waste Confidence update in 2010 that found that a repository would be available "when necessary" and that waste could be safely stored at reactor sites for at least 60 years after shutdown.[47]

The state of New York, environmental groups, and others filed lawsuits to overturn the 2010 Waste Confidence update on the grounds that NRC had not adequately considered the environmental risks of long-term waste storage at reactor sites. The U.S. Court of Appeals for the District of Columbia Circuit largely agreed, ruling on June 8, 2012, that NRC would have to conduct an environmental review of the Waste Confidence Decision under the National Environmental Policy Act (NEPA). The court found two major flaws in NRC's rulemaking process:

> First, in concluding that permanent storage will be available "when necessary," the Commission did not calculate the environmental effects of failing to secure permanent storage—a possibility that cannot be ignored. Second, in determining that spent fuel can safely be stored on site at nuclear

> plants for sixty years after the expiration of a plant's license, the Commission failed to properly examine future dangers and key consequences.[48]

NRC has halted licensing of new facilities that rely on the Waste Confidence Decision and plans to complete an environmental impact statement and issue a new Waste Confidence rule by September 2014.[49] The NRC commissioners approved a proposed Waste Confidence rule on August 5, 2013.[50]

CONGRESSIONAL ACTION

President Obama's proposal to terminate the Yucca Mountain project and search for disposal alternatives has prompted substantial congressional debate and a number of legislative proposals, with the discussion intensified by the release of DOE's January 2013 nuclear waste strategy. The House has consistently opposed the Obama Administration's efforts to abandon Yucca Mountain, while the Senate has generally expressed more interest in alternative waste management proposals.

Senator Wyden, along with Senators Murkowski, Feinstein, and Alexander, introduced legislation June 27, 2013, to redirect the nuclear waste program along the lines recommended by the Blue Ribbon Commission (S. 1240). The bill would establish an independent Nuclear Waste Administration to develop nuclear waste storage and disposal facilities. Siting of such facilities would require the consent of the affected state, local, and tribal governments.

The Nuclear Waste Administration would be required to develop a mission plan for opening a pilot storage facility for nuclear waste from shutdown reactors and other emergency deliveries by the end of 2021. Alarger storage facility for all other highly radioactive waste would have an operational target of December 31, 2025, and a permanent repository would have to be scheduled to open by the end of 2048 (the same dates as in the DOE strategy). The current disposal limit of 70,000 metric tons for the nation's first permanent repository would be repealed.

Nuclear waste fees collected after enactment of the bill would be held in a newly established Working Capital Fund. The Nuclear Waste Administration could immediately draw from that fund any amounts needed to carry out S. 1240, unless limited by annual appropriations or authorizations. Receipts and disbursements of waste fees, as well as the level of the fee, would be overseen by a Nuclear Waste Oversight Board appointed by the President with Senate confirmation. Fee collection would halt after 2025 if a waste facility had not been opened. The existing Nuclear Waste Fund would be available for appropriation by Congress to the Nuclear Waste Administration.

The Senate Committee on Energy and Natural Resources held a hearing on S. 1240 on July 30, 2013. Energy Secretary Ernest J. Moniz, a member of the Blue Ribbon Commission, said the bill provided "a promising framework for addressing key issues." NARUC Electricity Committee Chairman David C. Boyd called the bill "a step in the right direction," but urged that it require continued licensing action on the Yucca Mountain repository. Boyd noted that S. 1240 would not preclude enforcement of existing NWPA deadlines for action on Yucca Mountain. Natural Resources Defense Council Senior Attorney Geoffrey H. Fettus opposed the bill on the grounds that it would allow temporary waste storage facilities to be opened without progress on a permanent repository and that states would have inadequate authority to regulate repository safety, among other concerns.[51]

A pilot program to develop one or more voluntary interim storage sites for nuclear waste was included in the FY2014 Energy and Water Development appropriations bill passed by the Senate Appropriations Committee June 27, 2013 (S. 1245, §309). Before moving forward with any proposed waste storage facility, DOE would have to reach siting agreements with the governor of the proposed host state, each unit of local government with jurisdiction over the facility, and each affected Indian tribe. Such agreements would have to be approved by Congress, which would then appropriate funding for the project. Similar language was included by the committee in its FY2013 measure (S. 2465, §312). Corresponding House appropriations bills have not included such an authorization, and it has not been enacted.

The debate over nuclear waste policy was strongly affected by the March 11, 2011, Fukushima Daiichi nuclear accident in Japan. The loss of power at the Fukushima site, caused by a huge earthquake and tsunami, disabled cooling systems at the plant's spent fuel pools. Water in the pools was initially suspected to have boiled or leaked and dropped below the level of the stored spent fuel, potentially leading to fuel damage and radioactive releases into the atmosphere. However, later analysis indicated that the spent fuel did not overheat.

Concerns have been raised in Congress about the risk posed by stored spent fuel, particularly that cancellation of the Yucca Mountain repository would leave growing amounts of spent fuel indefinitely stored at nuclear plant sites throughout the United States. Legislation introduced by Representative

Engel on October 28, 2013, would require spent fuel at nuclear power plants to be moved from pools to dry casks after it has sufficiently cooled (H.R. 3354). Costs of the expedited spent fuel transfers would be offset by lower payments into the Nuclear Waste Fund. NRC released a study on November 12, 2013, concluding that "expedited transfer of spent fuel to dry cask storage would provide only a minor or limited safety benefit" and "its expected implementation costs would not be warranted."[52]

Nuclear waste policy was also the subject of numerous hearings in the 112th Congress, in both appropriations and authorizing committees. In a hearing June 1, 2011, Representative Shimkus, chairman of the Subcommittee on Environment and the Economy of the House Committee on Energy and Commerce, criticized DOE for halting the Yucca Mountain project without citing a technical basis for the decision. "Politics, not science, is driving the debate," he said in his opening statement.[53] At a June 14, 2011, hearing, the subcommittee questioned NRC Chairman Gregory Jaczko's decision to terminate the Yucca Mountain license review during FY2011, including the nearly completed safety evaluation report by the NRC staff.[54] After examining NRC's unreleased draft safety review, the majority staff of the House Committee on Science, Space, and Technology reported in June 2011 that NRC staff had "made over 1500 findings related to the scientific and technical research efforts of the Department of Energy" and that 98.5% of those findings had been in agreement with DOE's conclusions that the license should be issued.[55] As noted above, NRC has responded to a recent court decision by ordering its staff to complete the Yucca Mountain safety evaluation report with previously appropriated funds.

CHARACTERISTICS AND HANDLING OF NUCLEAR WASTE

Radioactive waste is a term that encompasses a broad range of material with widely varying characteristics. Some waste has relatively slight radioactivity and is safe to handle in unshielded containers, while other types are intensely hot in both temperature and radioactivity. Some decays to safe levels of radioactivity in a matter of days or weeks, while other types will remain dangerous for thousands of years. Major types of radioactive waste are described below:[56]

Spent nuclear fuel. Fuel rods that have been withdrawn from a nuclear reactor after irradiation, usually because they can no longer efficiently sustain a nuclear chain reaction. (The term "spent nuclear fuel" is defined in NWPA. The nuclear industry typically refers to spent fuel as "used nuclear fuel," because it contains uranium and plutonium that could be extracted through reprocessing to make new fuel.) By far the most radioactive type of civilian nuclear waste, spent fuel contains extremely hot but relatively short-lived fission products (fragments of the nuclei of uranium and other fissile elements) as well as long-lived radionuclides (radioactive atoms) such as plutonium, which remains dangerously radioactive for tens of thousands of years or more.

High-level waste. Highly radioactive residue created by spent fuel reprocessing (almost entirely for defense purposes in the United States). High-level waste contains most of the radioactive fission products of spent fuel, but most of the uranium and plutonium usually has been removed for re-use. Enough long-lived radioactive elements typically remain, however, to require isolation for 10,000 years or more.

Transuranic (TRU) waste. Relatively low-activity waste that contains more than a certain level of long-lived elements heavier than uranium (primarily plutonium). Radiation shielding may be required for handling of some types of TRU waste. In the United States, transuranic waste is generated almost entirely by nuclear weapons production processes. Because of the plutonium, long-term isolation is required. TRU waste is being sent to a deep underground repository, the Waste Isolation Pilot Plant (WIPP), near Carlsbad, NM.

Low-level waste. Radioactive waste not classified as spent fuel, high-level waste, TRU waste, or byproduct material such as uranium mill tailings (below). Four classes of low-level waste have been established by NRC, ranging from least radioactive and shortest-lived to the longest-lived and most radioactive. Although some types of low-level waste can be more radioactive than some types of high-level waste, in general low-level waste contains relatively low amounts of radioactivity that decays relatively quickly. Low-level waste disposal facilities cannot accept material that exceeds NRC concentration limits.

Uranium mill tailings. Sand-like residues remaining from the processing of uranium ore. Such tailings have very low radioactivity but extremely large volumes that can pose a hazard, particularly from radon emissions or groundwater contamination.

Mixed waste. Chemically hazardous waste that includes radioactive material. High-level, low-level, and TRU waste, and radioactive byproduct material, often falls under the designation of mixed waste. Such waste poses complicated institutional problems, because the radioactive portion is regulated by DOE or NRC under the Atomic EnergyAct, while the Environmental Protection Agency (EPA) and states regulate the non-radioactive elements under the Resource Conservation and RecoveryAct (RCRA).

Spent Nuclear Fuel

When spent nuclear fuel is removed from a reactor, usually after several years of power production, it is thermally hot and highly radioactive. The spent fuel is in the form of fuel assemblies, which consist of arrays of metal-clad fuel rods 12-15 feet long.

A fresh fuel rod, which emits relatively little radioactivity, contains uranium that has been enriched in the isotope U-235 (usually to 3%-5% from its natural level of 0.7%). But after nuclear fission has taken place in the reactor, most of the U-235 nuclei in the fuel rods have been split into a variety of highly radioactive fission products. Some of the nuclei of the dominant isotope U-238 have absorbed neutrons and then decayed to become radioactive plutonium, some of which has also split into fission products. Radioactive gases are contained in the spent fuel rods as well. Newly withdrawn spent fuel assemblies are stored in deep pools of water adjacent to the reactors to keep them from overheating and to protect workers from radiation. To prevent the pools from filling up, older, cooler spent fuel often is sealed in dry canisters and transferred to radiation-shielded storage facilities elsewhere at reactor sites. NRC currently requires spent fuel to cool for at least 7-10 years before being transferred to dry storage.[57]

Spent fuel discharged from U.S. commercial nuclear reactors is currently stored at 61 operating nuclear plant sites, 13 shutdown plant sites, and the Idaho National Laboratory.[58] Atypical large commercial nuclear reactor discharges an average of 20-30 metric tons of spent fuel per year—an average of about 2,150 metric tons annually for the entire U.S. nuclear power industry. The nuclear industry estimated that the total amount of commercial spent fuel was 69,718 metric tons at the end of 2012, including 20,104 metric tons in dry storage and other separate storage facilities.[59] Counting 7,000 metric tons of DOE spent fuel and high-level waste that had also been planned for disposal at Yucca Mountain, the total amount of existing waste would exceed NWPA's 70,000-metric-ton limit for the repository.

As long as nuclear power continues to be generated, the amount of spent fuel stored at plant sites will continue to grow until an interim storage facility or a permanent repository can be opened— or until alternative treatment and disposal technology is developed. DOE's most recent estimates of the total amount of U.S. commercial spent fuel that may eventually require disposal range from 105,000 metric tons[60] to 130,000 metric tons.[61]

New storage capacity at operating nuclear plant sites or other locations will be required if DOE is unable to begin accepting waste into its disposal system for an indefinite period. Most utilities are expected to construct new dry storage capacity at reactor sites. Fifty-six licensed dry storage facilities are currently operating in the United States.[62] NRC has determined that spent fuel could be stored safely at reactor sites for at least 60 years after a site's reactors cease operation (for a total of 120 years, assuming reactors are licensed for 60 years).[63] However, as noted above, that "waste confidence" determination is currently being revisited.

The terrorist attacks of September 11, 2001, heightened concerns about the vulnerability of stored spent fuel. Concerns have been raised that an aircraft crash into a reactor's pool area or acts of sabotage could drain the pool and cause the spent fuel inside to overheat. A report released by NRC January 17, 2001, found that overheating could cause the zirconium alloy cladding of spent fuel to catch fire and release hazardous amounts of radioactivity, although it characterized the probability of such a fire as low.

In a report released April 6, 2005, the National Academy of Sciences (NAS) found that "successful terrorist attacks on spent fuel pools, though difficult, are possible." To reduce the likelihood of spent fuel cladding fires, the NAS study recommended that hotter and cooler spent fuel assemblies be interspersed throughout spent fuel pools, that spray systems be installed above the pools, and that more fuel be transferred from pools to dry cask storage.[64] The nuclear industry contends that the several hours required for uncovered spent fuel to heat up enough to catch fire would allow ample time for alternative measures to cool the fuel.

As noted above, the Fukushima accident heightened concerns that spent fuel pools could be vulnerable to accidental damage resulting from the loss of cooling systems. The safety of spent fuel pools is one of the areas examined by an NRC task force that identified near-term lessons that the Fukushima accident may hold for U.S. nuclear power plant regulation. The task force recommended that assured sources of electrical power as well as water spray systems be available for spent fuel pools.[65] NRC approved an order March 9, 2012, requiring U.S. reactors to install improved water-level monitoring equipment at their spent fuel pools.[66] As noted above, NRC released a study on

November 12, 2013, that found only minor safety benefits in expedited transfers of spent fuel from pools to dry casks.[67]

Commercial Low-Level Waste

About 1.7 million cubic feet of low-level waste with about 9,400 curies of radioactivity was shipped to commercial disposal sites in 2012, according to DOE.[68] Volumes and radioactivity can vary widely from year to year, based on the status of nuclear decommissioning projects and cleanup activities that can generate especially large quantities. The radioactivity of low-level waste is only a tiny fraction of the amount in annual discharges of spent fuel.

Low-level radioactive waste is divided into three major categories for handling and disposal: Class A, B, and C. Classes B and C have constituted less than 1% of the volume of U.S. low-level waste disposal during the past five years but contain most of its radioactivity. As discussed below, most of the nation's Class B and C waste has been stored where it has been generated since June 2008 for lack of a permanent disposal site. The official opening of a new disposal facility for Class A, B, and C low-level waste near Andrews, TX, may alleviate the problem. Although the facility is intended to serve primarily Texas and Vermont, limited amounts of waste may be accepted from other states.[69]

CURRENT POLICY AND REGULATION

Disposal of spent fuel and high-level waste is a federal responsibility, while states are authorized to develop disposal facilities for commercial low-level waste. The Obama Administration has halted the Yucca Mountain project for high-level waste, although it remains the sole candidate repository site under current law. DOE issued a new waste management strategy in January 2013 that calls for a pilot facility for spent fuel storage to open at a voluntary site by 2021 and a new repository by 2048. New legislation would be required to carry out the strategy.

Spent Nuclear Fuel

Current Program and Proposed Policy Changes

The Nuclear Waste Policy Act established a system for selecting a geologic repository for the permanent disposal of up to 70,000 metric tons (77,000 tons) of spent nuclear fuel and high-level waste. DOE's Office of Civilian Radioactive Waste Management (OCRWM) was created to carry out the program. The Nuclear Waste Fund, holding receipts from a fee on commercial nuclear power) and federal contributions for emplacement of high-level defense waste, was established to pay for the program. The fee, set at a tenth of a cent per kilowatt-hour, can be adjusted by the Secretary of Energy based on projected total program costs after a congressional review period. DOE was required to select three candidate sites for the first national high-level waste repository.

After much controversy over DOE's implementation of NWPA, the act was substantially modified by the Nuclear Waste Policy Amendments Act of 1987 (Title IV, Subtitle A of P.L. 100-203, the Omnibus Budget Reconciliation Act of 1987). Under the amendments, the only candidate site DOE may consider for a permanent high-level waste repository is at Yucca Mountain, Nevada. If that site cannot be licensed, DOE must return to Congress for further instructions.

The 1987 amendments also authorized construction of a monitored retrievable storage (MRS) facility to store spent fuel and prepare it for delivery to the repository. But because of fears that the MRS would reduce the need to open the permanent repository and become a de facto repository itself, the law forbids DOE from selecting an MRS site until recommending to the President that a permanent repository be constructed, and construction of an MRS cannot begin until Yucca Mountain receives a construction permit. The repository recommendation was made in February 2002, but DOE has not announced any plans for siting an MRS.

Along with halting all funding for the Yucca Mountain project, the Obama Administration terminated OCRWM at the end of FY2010 and transferred its remaining functions to DOE's Office of Nuclear Energy. The Administration established the Blue Ribbon Commission on America's Nuclear Future (BRC) to develop a new waste management strategy, and the BRC issued its final report on January 26, 2012.[70]

As required by its charter, the BRC did not evaluate specific sites for new nuclear waste facilities, including Yucca Mountain. However, the commission concluded that the existing nuclear waste policy, with Yucca Mountain identified by law as the sole candidate site, "has now all but completely broken down" and "seems destined to bring further controversy, litigation, and protracted delay." The BRC recommended instead that Congress establish "a new, consent-based approach to siting." Under that approach, potential sites would be the subject of extensive negotiations with affected states, tribes, and local governments. Such negotiations would result in legally binding agreements on the roles of the affected parties, including local oversight, and other project parameters.

The BRC noted that previous U.S. efforts to find voluntary waste sites had failed, but it nevertheless expressed confidence that such a process could eventually succeed. In particular, the commission highlighted the U.S. experience with the Waste Isolation Pilot Plant (WIPP) in New Mexico, which, after many years of controversy, began receiving transuranic defense waste in 1999 with state and local government approval.

Local officials near the WIPP facility have long supported the development of additional waste facilities at the site, which was originally planned to hold high-level waste before the state objected. A presentation by a top New Mexico official on March 1, 2012, described conditions under which the state might be willing to accept high-level waste and spent fuel at the WIPP site, such as assistance with cleaning up the state's contaminated uranium production sites.[71] A local government consortium near the WIPP site, the Eddy-Lea Energy Alliance, sent a letter to NRC on February 26, 2013, that it would submit a license application for the consolidated spent fuel storage facility envisioned by DOE's waste strategy report. "As details of the DOE strategy are implemented, we will keep the NRC staff advised of our progress," the letter said.[72] Interest in hosting nuclear waste sites has also been expressed by groups in Mississippi and Loving County, Texas.[73]

To carry out the new waste management program, the BRC recommended that a congressionally chartered federal corporation be established. Such a corporation would be independent from Administration control and have "assured access to funds" but be subject to congressional oversight and to regulation by NRC. Pending establishment of the corporation, the BRC recommended that administrative and legislative changes be implemented in the Nuclear Waste Fund to allow funds to be used for the waste management program without having to compete with other budget priorities.

The BRC called for "prompt efforts" to develop a permanent underground nuclear waste repository and to develop one or more interim central storage facilities. Interim storage facilities are especially needed so that waste can be removed from shutdown reactor sites, the commission said. Development of a permanent site would have to be undertaken along with the interim storage effort to assure that interim sites would not become "de facto" permanent repositories, according to the commission.

In response to the BRC report, and to provide an outline for a new nuclear waste program, DOE issued its *Strategy for the Management and Disposal of Used Nuclear Fuel and High-Level Waste* in January 2013.[74] Under the DOE strategy, a pilot interim spent fuel storage facility would be opened by 2021, focusing primarily on spent fuel from decommissioned nuclear plants. A larger-scale interim storage facility, which could be an expansion of the pilot facility, would open by 2025 with a capacity of 20,000 metric tons or more.

The DOE strategy called for the interim storage facility to be linked to development of a permanent repository so that the storage facility would not become a *de facto* repository. However, the strategy noted that the existing NWPA restrictions on the MRS are so rigid that the MRS cannot currently be built. Without describing specific provisions, the DOE strategy recommended that "this linkage should not be such that it overly restricts forward movement on a pilot or larger storage facility that could make progress against the waste management mission."

Under the DOE strategy, a geologic disposal facility would open by 2048—50 years after the initially planned opening date for the Yucca Mountain repository. A site for the repository is to be selected by 2016, and site suitability studies, design, and licensing are to be completed by 2042.

Sites for the proposed storage and disposal facilities would be selected through a "consent based" process, as recommended by the BRC. However, the DOE strategy included few details on how such a process would be implemented. Instead, the strategy said the Obama Administration would soon begin consultations with Congress and interest groups on "defining consent, deciding how that consent is codified, and determining whether or how it is ratified by Congress." The program would be implemented by a new nuclear waste management entity, as recommended by the BRC, but the nature of the new organization was not specified by the DOE strategy. A bill introduced by Senator Wyden (S. 1240), discussed under "Congressional Action," would largely authorize the Administration's new waste strategy.

Private Interim Storage

When it became apparent that DOE would miss the 1998 deadline for taking nuclear waste from reactor sites, a utility consortium signed an agreement with the Skull Valley Band of the Goshute Indians in Utah on

December 27, 1996, to develop a private spent fuel storage facility on tribal land. The Private Fuel Storage (PFS) consortium submitted a license application to NRC on June 25, 1997, and a 20-year license for storing up to 44,000 tons of spent fuel in dry casks was issued on February 21, 2006, although NRC noted that Interior Department approval would also be required.

On September 7, 2006, the Department of the Interior issued two decisions against the PFS project. The Bureau of Indian Affairs disapproved a proposed lease of tribal trust lands to PFS, concluding there was too much risk that the waste could remain at the site indefinitely.[75] The Bureau of Land Management rejected the necessary rights-of-way to transport waste to the facility, concluding that a proposed rail line would be incompatible with the Cedar Mountain Wilderness Area and that existing roads would be inadequate.[76]

The Skull Valley Band of Goshutes and PFS filed a federal lawsuit July 17, 2007, to overturn the Interior decisions on the grounds that they were politically motivated.[77] A federal district court judge on July 26, 2010, ordered the Department of the Interior to reconsider its decisions on the PFS permits.[78] However, PFS asked NRC to terminate its license on December 20, 2012.[79]

Regulatory Requirements for Yucca Mountain

Although the Obama Administration wants to redirect the nuclear waste program, current law still focuses on Yucca Mountain. NWPA requires that high-level waste repositories be licensed by NRC in accordance with general standards issued by EPA. Under the Energy Policy Act of 1992 (P.L. 102-486), EPA was required to write new repository standards specifically for Yucca Mountain. NWPA also requires the repository to meet general siting guidelines prepared by DOE and approved by NRC. Transportation of waste to storage and disposal sites is regulated by NRC and the Department of Transportation (DOT). Under NWPA, DOE shipments to Yucca Mountain and an MRS facility would have to use NRC-certified casks and comply with NRC requirements for notifying state and local governments. Shipments would also have to follow DOT regulations on routing, placarding, and safety.

NRC's licensing requirements for Yucca Mountain, at 10 C.F.R. 63, require compliance with EPA's standards (described below) and establish procedures that DOE must follow in seeking a repository license. For example, DOE is required to conduct a repository performance confirmation program that would indicate whether natural and man-made systems were functioning as intended and assure that other assumptions about repository conditions were accurate.

Specific standards for Yucca Mountain were required because of concerns that some of EPA's general standards might be impossible or impractical to meet at Yucca Mountain.[80] The new standards, which limit the radiation dose that the repository could impose on individual members of the public, were required to be consistent with the findings of a study by the National Academy of Sciences (NAS), which was issued August 1, 1995.[81] The NAS study recommended that the Yucca Mountain environmental standards establish a limit on risk to individuals near the repository, rather than setting specific limits for the releases of radioactive material or on radioactive doses, as under previous EPA standards. The NAS study also examined the potential for human intrusion into the repository and found no scientific basis for predicting human behavior thousands of years into the future.

Pursuant to the Energy Policy Act of 1992, EPA published its proposed Yucca Mountain radiation protection standards on August 27, 1999. The proposal would have limited annual radiation doses to 15 millirems for the "reasonably maximally exposed individual," and to 4 millirems from groundwater exposure, for the first 10,000 years of repository operation. EPA calculated that its standard would result in an annual risk of fatal cancer for the maximally exposed individual of seven chances in a million. The nuclear industry criticized the EPA proposal as being unnecessarily stringent, particularly the groundwater standard. On the other hand, environmental groups contended that the 10,000-year standard proposed by EPA was too short, because DOE had projected that radioactive releases from the repository would peak after about 400,000 years.

EPA issued its final Yucca Mountain standards on June 6, 2001. The final standards included most of the major provisions of the proposed version, including the 15 millirem overall exposure limit and the 4 millirem groundwater limit. Despite the department's opposition to the EPA standards, DOE's site suitability evaluation determined that the Yucca Mountain site would be able to meet them. NRC revised its repository regulations September 7, 2001, to conform to the EPA standards.

A three-judge U.S. Court of Appeals panel on July 9, 2004, struck down the 10,000-year regulatory compliance period in the EPA and NRC Yucca Mountain standards.[82] The court ruled that the 10,000-year period was inconsistent with the NAS study on which the Energy Policy Act required the Yucca Mountain regulations to be based. In fact, the court found, the NAS study had specifically rejected a 10,000-year compliance period because of analysis that showed peak radioactive exposures from the repository would take place several hundred thousand years in the future.

In response to the court decision, EPA proposed a new version of the Yucca Mountain standards on August 9, 2005. The proposal would have retained the dose limits of the previous standard for the first 10,000 years but allowed a higher annual dose of 350 millirems for the period of 10,000 years through 1 million years. EPA also proposed to base the post-10,000-year Yucca Mountain standard on the median dose, rather than the mean, potentially making it easier to meet.[83] Nevada state officials called EPA's proposed standard far too lenient and charged that it was "unlawful and arbitrary."[84]

EPA issued its final rule to amend the Yucca Mountain standards on September 30, 2008. The final rule reduced the annual dose limit during the period of 10,000 through 1 million years from the proposed 350 millirems to 100 millirems, which the agency contended was consistent with international standards. Under the final rule, compliance with the post-10,000-year standard will be based on the arithmetic mean of projected doses, rather than the median as proposed. The 4 millirem groundwater standard will continue to apply only to the first 10,000 years.[85] NRC revised its repository licensing regulations to conform to the new EPA standards onApril 13, 2009.[86]

DOE estimated in its June 2008 Final Supplemental Environmental Impact Statement (FSEIS) for the Yucca Mountain repository that the maximum mean annual individual dose after 10,000 years would be 2 millirems. That is substantially below the level estimated by the 2002 Final Environmental Impact Statement, which calculated that the peak doses—occurring after 400,000 years—would be about 150 millirems (Volume 1, Chapter 5). The FSEIS attributed the reduction to changes in DOE's computer model and in the assumptions used, noting that "various elements of DOE's modeling approach may be challenged as part of the NRC licensing process."[87]

Alternative Technologies

Several alternatives to the geologic disposal of spent fuel have been studied by DOE and its predecessor agencies, as well as technologies that might reduce waste disposal risks. However, most of these technologies involve large technical obstacles, uncertain costs, and potential public opposition.

Among the primary long-term disposal alternatives to geologic repositories are disposal below the seabed and transport into space, neither of which is currently being studied by DOE. Other technologies have been studied that, while probably not replacing geologic disposal, might make geologic disposal safer and more predictable. Chief among these is the reprocessing or "recycling" of spent fuel so that plutonium, uranium, and other long-lived radionuclides could be converted to faster-decaying fission products in special nuclear reactors or particle accelerators. Emplacing waste in deep boreholes, at much greater depths than most proposed repositories, has also been suggested.

DOE's Fuel Cycle Research and Development Program conducts "long-term, science-based" research on a wide variety of technologies for improving the management of spent nuclear fuel, according to the DOE budget justification. The total FY2014 funding request for this program was $165.1 million, down from the FY2013 level of $187.4 million.

Under the Obama Administration, the program has been redirected toward development of technology options for a wide range of nuclear fuel cycle approaches, including direct disposal of spent fuel (the "once through" cycle) and partial and full recycling, according to the justification. The fuel cycle R&D program "will research and develop a suite of technology options that will enable future decision-makers to make informed decisions about how best to manage nuclear waste and used fuel from reactors," the FY2014 justification says.

Funding

The Obama Administration's FY2011 budget request called for a complete halt in funding for the Yucca Mountain project and elimination of OCRWM. In line with the request, the FY2011 Continuing Appropriations Act (P.L. 112-10) provided no DOE funding for the program. DOE shut down the Yucca Mountain project at the end of FY2010 and transferred OCRWM's remaining functions to the Office of Nuclear Energy.

President Obama's FY2014 federal budget, submitted to Congress April 11, 2013, included no funding for Yucca Mountain but requested $60 million "to support the program of work" outlined by DOE's January 2013 nuclear waste strategy. The House provided none of the requested nuclear waste funding in the FY2014 Energy and Water Development appropriations bill (H.R. 2609, H.Rept. 113-135), passed July 10, 2013. Instead, the House included $25 million in the bill for work related to Yucca Mountain. The Senate Appropriations Committee endorsed the Administration's nuclear waste funding request in its version of the Energy and Water Bill (S. 1245, S. Rept. 113-47), approved June 27, 2013. FY2014 nuclear waste funding is currently provided through January 15, 2014, by a continuing resolution (P.L.

113-46), at about the spending rate requested by the Administration and with no new funds for Yucca Mountain.

Although nuclear utilities pay fees to the Nuclear Waste Fund to cover the disposal costs of civilian nuclear spent fuel, DOE cannot spend the money in the fund until it is appropriated by Congress. At the end of FY2012, the Waste Fund balance stood at $28.7 billion, according to the Treasury Department.[88] Before the Obama Administration halted the Yucca Mountain project, $7.41 billion had been disbursed from the Waste Fund, according to DOE's program summary report.[89] DOE's most recent update of its *Analysis of the Total System Life Cycle Cost of the Civilian Radioactive Waste Management Program* was released on August 5, 2008.[90] According to that estimate, the Yucca Mountain program as then planned would cost $96.2 billion in 2007 dollars from the beginning of the program in 1983 to repository closure in 2133.

Low-Level Radioactive Waste

Current Policy

Selecting disposal sites for low-level radioactive waste, which generally consists of low concentrations of relatively short-lived radionuclides, is authorized to be conducted by states under the 1980 Low-Level Radioactive Waste Policy Act and 1985 amendments. Most states have joined congressionally approved interstate compacts to handle low-level waste disposal. Under the 1985 amendments, the nation's three (at that time) operating commercial low-level waste disposal facilities could start refusing to accept waste from outside their regional interstate compacts after the end of 1992. One of the three sites, near Beatty, NV, closed. The remaining two—at Barnwell, SC, and Hanford, WA—are using their congressionally granted authority to prohibit waste from outside their regional compacts. Another site, in Utah, has since become available nationwide for most Class A low-level waste. A site in Texas, discussed below, is authorized to import class A, B, and C waste, but until those imports begin, most of the nation's class B and C waste is being stored at the sites where it is generated.

The commercial low-level waste disposal site under the jurisdiction of the Texas Compact, consisting of Texas and Vermont, opened on November 10, 2011.[91] The site received its first shipment of waste, from a company in Vermont, on April 27, 2012.[92] Texas also enacted a law in 2011 to allow up to

30% of the Texas disposal site's capacity to be used by states outside of the compact, and final regulations for the waste imports were issued March 23, 2012.[93]

Legislation providing congressional consent to the Texas compact, which originally also included Maine as well as Vermont, was signed by President Clinton September 20, 1998 (P.L. 105-236). However, on October 22, 1998, a proposed disposal site near Sierra Blanca, TX, was rejected by the Texas Natural Resource Conservation Commission, and Maine subsequently withdrew. Texas Governor Perry signed legislation June 20, 2003, authorizing the Texas Commission on Environmental Quality (TCEQ) to license adjoining disposal facilities for commercial and federally generated low-level waste. Pursuant to that statute, an application to build the Andrews County disposal facility was filed August 2, 2004, by Waste Control Specialists LLC. TCEQ voted January 14, 2009, to issue the license after the necessary land and mineral rights had been acquired and approved construction of the facility January 7, 2011.[94]

The disposal facility at Barnwell, SC, is currently accepting all ClassA, B, and C low-level waste from the Atlantic Compact (formerly the Northeast Compact), in which South Carolina joined original members Connecticut and New Jersey on July 1, 2000. Under the compact, South Carolina can limit the use of the Barnwell facility to the three compact members, and a state law enacted in June 2000 phased out acceptance of non-compact waste through June 30, 2008. The Barnwell facility previously had stopped accepting waste from outside the Southeast Compact at the end of June 1994. The Southeast Compact Commission in May 1995 twice rejected a South Carolina proposal to open the Barnwell site to waste generators outside the Southeast and to bar access to North Carolina until that state opened a new regional disposal facility, as required by the compact. The rejection of those proposals led the South Carolina General Assembly to vote in 1995 to withdraw from the Southeast Compact and begin accepting waste at Barnwell from all states but North Carolina. North Carolina withdrew from the Southeast Compact July 26, 1999. The U.S. Supreme Court ruled on June 1, 2010, that the withdrawal did not subject North Carolina to sanctions under the compact.[95]

The only other existing disposal facility for all three major classes of low-level waste is at Hanford, WA. Controlled by the Northwest Compact, the Hanford site will continue taking waste from the neighboring Rocky Mountain Compact under a contract. Since the South Carolina facility closed to out-of-region waste, and pending planned imports by the Texas compact, the 34 states and the District of Columbia that are outside the Northwest, Rocky Mountain, Atlantic, and Texas compacts have had no disposal site for Class B and C low-level waste.

Regulatory Requirements

Licensing of commercial low-level waste facilities is carried out under the Atomic Energy Act by NRC or by "agreement states" with regulatory programs approved by NRC. NRC regulations governing low-level waste licenses must conform to general environmental protection standards and radiation protection guidelines issued by EPA. Transportation of low-level waste is jointly regulated by NRC and the Department of Transportation.

CONCLUDING DISCUSSION

Disposal of radioactive waste will be a key issue in the continuing nuclear power debate. Without a national disposal system, spent fuel from nuclear power plants must be stored on-site indefinitely. This situation may raise public concern near proposed reactor sites, particularly at sites without existing reactors where spent nuclear fuel is already stored. Concern about spent fuel storage safety has been heightened by the March 2011 disaster at Japan's Fukushima Daiichi nuclear plant.

Under current law, the federal government's nuclear waste disposal policy is focused on the Yucca Mountain site. However, President Obama's moves to terminate the Yucca Mountain project and develop a new waste strategy through the Blue Ribbon Commission on America's Nuclear Future have brought most activities in the DOE waste program to a halt. Congress is continuing to debate the project's termination, particularly through the appropriations process.

Because of their waste-disposal contracts with DOE, owners of existing reactors are likely to continue seeking damages from the federal government if disposal delays continue. DOE's 2004 settlement with the nation's largest nuclear operator, Exelon, could require payments of up to $600 million from the federal judgment fund, for example. DOE estimates that payments could rise above $20 billion if the federal government cannot begin taking waste from reactor sites before 2020, as previously planned. The nuclear industry has predicted that future damages could rise by tens of billions of dollars if the federal disposal program fails altogether.

Lack of a nuclear waste disposal system could also affect the licensing of proposed new nuclear plants, both because of NRC licensing guidelines and various state laws.[96] In addition, further repository delays could force DOE to miss compliance deadlines for defense waste disposal.

Problems being created by nuclear waste disposal delays were addressed by the Blue Ribbon Commission in its final report, issued in January 2012. Major options include centralized interim storage, continued storage at existing nuclear sites, reprocessing and waste treatment technology, development of alternative repository sites, or a combination. The commission recommended that a congressionally chartered corporation be established to undertake a negotiated process for siting new waste storage and disposal facilities. However, given the delays resulting from the ongoing shutdown of the nuclear waste program, longer on-site storage is almost a certainty under any option. Any of the options would also face intense controversy, especially among states and regions that might be potential hosts for future waste facilities. As a result, substantial debate would be expected over any proposals to change the Nuclear Waste Policy Act, including those of the Blue Ribbon Commission.

SELECTED LEGISLATION

H.R. 2081 (Thornberry)

No More Excuses Energy Act of 2013. Includes provisions to prohibit NRC from considering nuclear waste storage when licensing new nuclear facilities, and to establish a tax credit for obtaining nuclear component manufacturing certification. Introduced May 21, 2013; referred to multiple committees.

H.R. 2609 (Frelinghuysen)/S. 1245 (Feinstein)

Energy and Water Development and Related Agencies Appropriations Act, 2014. Provides funding for DOE nuclear programs and NRC. House bill introduced July 2, 2013; reported as original measure by Committee on Appropriations July 2, 2013 (H.Rept. 113-135); passed House July 10, 2013, by vote of 227-198. Senate bill introduced June 27, 2013; reported as original measure by Committee on Appropriations June 27, 2013 (S.Rept. 113-47).

H.R. 3354 (Engel)

Dry Cask Storage Act. Requires spent fuel at nuclear power plants to be moved from spent fuel pools to dry casks after it has sufficiently cooled. Costs of the fuel transfers would be offset by a reduction in nuclear waste fees owed to the federal government. Introduced October 28, 2013; referred to Committee on Energy and Commerce.

S. 1240 (Wyden)

Nuclear Waste Administration Act of 2013. Establishes an independent Nuclear Waste Administration to develop nuclear waste storage and disposal facilities. Siting of such facilities would require the consent of the affected state, local, and tribal governments. The Nuclear Waste Administration could spend nuclear waste fees collected after the bill's enactment without the need for further appropriation. Fee collection would halt after 2025 if a waste facility had not been opened. Introduced June 27, 2013; referred to Committee on Energy and Natural Resources. Committee hearing held July 30, 2013.

For Additional Reading

Blue Ribbon Commission onAmerica's Nuclear Future.

Report to the Secretary of Energy, January 2012, http://brc.gov/sites/default/files/documents/ brc_finalreport_jan2012.pdf.

Commissioned Papers. 2010-2011. Reports on current nuclear waste issues. http://brc.gov/index.php?q=library/documents/commissioned-papers.

Harvard University. John F. Kennedy School of Government. Belfer Center for Science and International Affairs. *The Economics of Reprocessing vs. Direct Disposal of Spent Nuclear Fuel.* DE-FG26-99FT4028. December 2003.

Nuclear Waste Technical Review Board.

*Experience Gained from Programs to Manage High-Level Radioactive Waste and Spent Nuclear Fuel in the United States and Other Countries.*April 2011. http://www.nwtrb.gov/ reports/reports.html.

Survey of National Programs for Managing High-Level Radioactive Waste and Spent Nuclear Fuel. October 2009. http://www.nwtrb.gov/reports/reports.html.

RAND Corporation. *Managing Spent Nuclear Fuel: Strategy Alternatives and Policy Implications*. 2010. 71 p. http://www.rand.org/content/dam/rand/pubs/monographs/2010/ RAND_MG970.pdf.

University of Illinois. Program in Arms Control, Disarmament, and International Security. *'Plan D' for Spent Nuclear Fuel.* 2009. http://acdis.illinois.edu/publications/207/publication-PlanDforSpentNuclearFuel.html.

U.S. Department of Energy.

Strategy for the Management and Disposal of Used Nuclear Fuel and High-Level Radioactive Waste, January 2013, http://energy.gov/sites/prod/files/Strategy%20for%20the%20Management%20and%20Disposal%20of%20Used%20Nuclear%20Fuel%20and%20High%20Level%20Radioactive%20Waste.pdf.

Used Fuel Disposition Campaign: Disposal Research and Development Roadmap, March 2011, http://energy.gov/sites/prod/files/UFD_Disposal_R%26 D_Roadmap_Rev_0.1.pdf.

Walker, J. Samuel. *The Road to Yucca Mountain: The Development of Radioactive Waste Policy in the United States.* University of California Press. 2009. 228 p.

End Notes

[1] DOE, Strategy for the Management and Disposal of Used Nuclear Fuel and High-Level Waste, January 2013, http://energy.gov/sites/prod/files/ Strategy%20for%20the%20Management%20and%20Disposal%20of%20Used%20Nuclear%20Fuel%20and%20High%20Level%20Radioactive%20Waste.pdf.

[2] See CRS Legal Sidebar WSLG734, *Court Neither Razzled Nor Dazzled by DOE's Failure to Assess Nuclear Waste Fund Fee*, by Todd Garvey.

[3] U.S. Circuit Court of Appeals for the District of Columbia Circuit, *State of New York, et al. v. Nuclear Regulatory Commission*, No. 11-1045, Decided June 8, 2012, http://www.cadc.uscourts.gov/internet/opinions.nsf/57ACA94A8FFAD8AF85257A1700502AA4/$file/11-1045-1377720.pdf.

[4] Nuclear Regulatory Commission, "Waste Confidence," web page, February 20, 2013, http://www.nrc.gov/waste/ spent-fuel-storage/wcd.html.

[5] NRC, "Proposed Rule: Waste Confidence," August 5, 2013, http://pbadupws.nrc.gov/docs/ML1321/ ML13217A358.pdf.

[6] Nuclear Regulatory Commission, "NRC to Issue Orders, Information Request as Part of Implementing Fukushima-Related Recommendations," press release, March 9, 2012, http://www.nrc.gov/reading-rm/doc-collections/news/2012/ 12-023.pdf.

[7] Nuclear Energy Institute, Key Issues, Yucca Mountain, http://www.nei.org/keyissues/ nuclearwastedisposal/yuccamountain/, viewed April 11, 2008.

[8] *U.S. Department of Energy's Motion to Withdraw*, NRC Atomic Safety and Licensing Board, Docket No. 63-0001, March 3, 2010, http://www.energy.gov/news/documents/ DOE_Motion_to_Withdraw.pdf.

[9] National Association of Regulatory Utility Commissioners, "NARUC Seeks Party Status at NRC, Says Yucca Review Must Continue," press release, March 16, 2010, http://www.naruc. org/News/default.cfm?pr=191&pdf=.

[10] U.S. Nuclear Regulatory Commission, Atomic Safety and Licensing Board, Docket No. 63-001-HLW, Memorandum and Order, June 29, 2010.

[11] Nuclear Regulatory Commission, "In the Matter of U.S. Department of Energy (High-Level Waste Repository)," CLI-11-07, September 9, 2011, http://www.nrc.gov/reading-rm/doc-collections/commission/orders/2011/2011-07cli.pdf.

[12] U.S. Court of Appeals for the District of Columbia Circuit, *In re: Aiken County et al.*, No. 11-1271, writ of mandamus, August 13, 2013, http://www.cadc.uscourts.gov/internet/ opinions.nsf/BAE0CF34F762EBD985257BC6004DEB18/$file/11-1271-1451347.pdf.

[13] Nuclear Regulatory Commission, "NRC Directs Staff to Complete Yucca Mountain Safety Evaluation Report," news release No. 13-097, November 18, 2013, http://pbadupws. nrc.gov/docs/ML1332/ML13322B228.pdf.

[14] Blue Ribbon Commission on America's Nuclear Future, *Report to the Secretary of Energy*, January 2012, http://brc.gov/sites/default/files/documents/brc_finalreport_jan2012.pdf (BRC Final Report).

[15] Blue Ribbon Commission on America's Nuclear Future, Commissioned Papers, http://brc.gov/index.php?q=library/ documents/commissioned-papers.

[16] DOE, *Strategy for the Management and Disposal of Used Nuclear Fuel and High-Level Waste*, op. cit.

[17] National Research Council, Board on Radioactive Waste Management, *Rethinking High-Level Radioactive Waste Disposal: A Position Statement of the Board on Radioactive Waste Management* (1990), p. 2.

[18] Senator Bingaman introduced the approval resolution in the Senate April 9, 2002 (S.J.Res. 34), and Representative Barton introduced it in the House April 11, 2002 (H.J.Res. 87). The Subcommittee on Energy and Air Quality of the House Committee on Energy and Commerce approved H.J.Res. 87 on April 23 by a 24-2 vote, and the full committee approved the measure two days later, 41-6 (H.Rept. 107-425). The resolution was passed by the House May 8, 2002, by a vote of 306-117. The Senate Committee on Energy and Natural Resources approved S.J.Res. 34 by a 13-10 vote June 5, 2002 (S.Rept. 107-159). Following a 60-39 vote to consider S.J.Res. 34, the Senate passed H.J.Res. 87 by voice vote July 9, 2002.

[19] Waste Control Specialists LLC, "Historic Texas Compact Disposal Facility Ready for Business," http://www.wcstexas.com.

[20] Waste Control Specialists LLC, "Waste Control Specialists Commends Passage of Legislation," press release, May 31, 2011, http://www.wcstexas.com/PDF_downloads/ WCSAnnounceslegislation.pdf?nxd_id=98546.

[21] The Standard Contract for Disposal of Spent Nuclear Fuel and/or High-Level Radioactive Waste can be found at 10 CFR 961.11.

[22] Derek Sands and Elaine Hiruo, "Piloting Spent Fuel Storage Is Path to Larger Facility: DOE Official," *Nucleonics Week*, January 17, 2013.

[23] Ernest J. Moniz, Secretary of Energy, Statement before the Senate Energy and Natural Resources Committee, July 30, 2013, http://energy.gov/sites/prod/files/2013/07/f2/7-30-13_Ernest_Moniz%20FT%20SENR.pdf.

[24] U.S. Department of Energy, Office of Civilian Radioactive Waste Management, *Analysis of the Total System Life Cycle Cost of the Civilian Radioactive Waste Management Program, Fiscal Year 2007*, DOE/RW-0591, Washington, DC, July 2008, p. 20, http://ocrwm.doe.gov/about/budget/pdf/TSLCC_2007_8_05_08.pdf.

[25] BRC Final Report, op. cit., p. 80.

[26] Ibid., p. B-2.

[27] NARUC, "State Regulators Go to Court with DOE over Nuclear Waste Fees, news release, April 2, 2010, http://www.naruc.org/News/default.cfm?pr=193; *Nuclear Energy Institute et al. v. U.S. DOE*, Joint Petition for Review, U.S. Court of Appeals for the District of Columbia Circuit, April 5, 2010.

[28] Jeff Beattie, "NARUC, Utilities Sue DOE Over Nuke Waste Fee," *Energy Daily*, April 6, 2010, p. 1.

[29] Secretary of Energy Steven Chu, "Secretarial Determination of the Adequacy of the Nuclear Waste Fund Fee," November 1, 2010, http://energy.gov/sites/prod/files/gcprod/documents/ Secretarial_Determination_WasteFee.pdf.

[30] U.S. Court of Appeals for the District of Columbia Circuit, *National Association of Regulatory Utility Commissioners v. United States Department of Energy*, No. 11-1066, decided June 1, 2012, http://www.cadc.uscourts.gov/internet/opinions.nsf/4B11622F4FF75FEC85257A 100050A681/$file/11-1066-1376508.pdf.

[31] DOE, "Nuclear Waste Fund Fee Adequacy Report," January 2013, http://energy.gov/sites/ prod/files/January%2016%202013%20Secretarial%20Determination%20of%20the%20Ade quacy%20of%20the%20Nuclear%20 Waste%20Fund%20Fee_0.pdf.

[32] See CRS Legal Sidebar WSLG734, *Court Neither Razzled Nor Dazzled by DOE's Failure to Assess Nuclear Waste Fund Fee*, by Todd Garvey.

[33] Statement of Peter B. Lyons, Assistant Secretary for Nuclear Energy, U.S. Department of Energy, before the Committee on Energy and Commerce, Subcommittee on Environment and the Economy, June 1, 2011.

[34] DOE Motion to Withdraw, op. cit.

[35] Nicole E. Matthews, "DOE Withdraws Application for Yucca Nuke Dump," *Fox5Vegas.com*, March 3, 2010, http://www.fox5vegas.com/news/22734591/detail.html.

[36] *Motion for the State of Nevada for Leave to Intervene as Intervenor-Respondent*, U.S. Court of Appeals for the Fourth Circuit, Case No. 10-1229, March 19, 2010, http://www. state.nv.us/nucwaste/licensing/nv100319motion3.pdf.

[37] National Association of Regulatory Utility Commissioners, "NARUC Seeks Party Status at NRC, Says Yucca Review Must Continue," press release, March 16, 2010, http:// www.naruc.org/News/default.cfm?pr=191&pdf=.

[38] U.S. Nuclear Regulatory Commission, Atomic Safety and Licensing Board, Docket No. 63-001-HLW, Memorandum and Order, June 29, 2010.

[39] *U.S. Department of Energy's Brief in Support of Review and Reversal of the Board's Ruling on the Motion to Withdraw*, Docket No. 63-001-HLW, July 9, 2010.

[40] Nuclear Regulatory Commission, Memorandum and Order, CLI-11-07, September 9, 2011, http://www.nrc.gov/ reading-rm/doc-collections/commission/orders/2011/2011-07cli.pdf.

[41] U.S. Circuit Court of Appeals for the District of Columbia Circuit, USCA Case #11-1271, Yucca Mountain Reply Brief of Petitioners Mandamus Action, February 13, 2012, http://www.naruc.org/policy.cfm?c=filings.

[42] Freebairn, William, "NRC Says Lack of Funding Supports Decision to Stop Yucca Review," *NuclearFuel*, January 21, 2013.

[43] U.S. Court of Appeals for the District of Columbia Circuit, In re: Aiken County et al., No. 11-1271, writ of mandamus, August 13, 2013, http://www.cadc.uscourts.gov/internet/ opinions.nsf/BAE0CF34F762EBD985257BC6004DEB18/$file/11-1271-1451347.pdf.

[44] U.S. Circuit Court of Appeals for the District of Columbia Circuit, *Minnesota v. NRC*, 602 F.2d 412 (D.C. Cir. 1979).

[45] NRC, "Waste Confidence Decision," 49 *Federal Register* 34,658, August 31, 1984.

[46] NRC, "Waste Confidence Decision Review," 55 *Federal Register* 38,474, September 18, 1990.

[47] NRC, "Waste Confidence Decision Update," 75 *Federal Register* 81,037, December 23, 2010.

[48] U.S. Circuit Court of Appeals for the District of Columbia Circuit, *State of New York, et al. v. Nuclear Regulatory Commission*, No. 11-1045, Decided June 8, 2012, http://www.cadc.uscourts.gov/internet/opinions.nsf/57ACA94A8FFAD8AF85257A1700502AA4/$file/11-1045-1377720.pdf.

[49] Nuclear Regulatory Commission, "Waste Confidence," web page, February 20, 2013, http://www.nrc.gov/waste/ spent-fuel-storage/wcd.html.

[50] NRC, "Proposed Rule: Waste Confidence," August 5, 2013, http://pbadupws.nrc.gov/docs/ML1321/ ML13217A358.pdf.

[51] Senate Committee on Energy and Natural Resources, Full Committee Hearing to Consider the Nuclear Waste Administration Act of 2013, July 30, 2013, http://www.energy.senate.gov/public/index.cfm/2013/7/full-committee-hearing-to-consider-the-nuclear-waste-administration-act-of-2013.

[52] Nuclear Regulatory Commission, "Staff Evaluation and Recommendations for Japan Lessons-Learned Tier 3 Issue on Expedited Transfer of Spent Fuel," COMSECY-13-0030, November 12, 2013, http://www.nrc.gov/reading-rm/doc-collections/commission/comm-secy/2013/2013-0030comscy.pdf.

[53] Opening Statement, Chairman John Shimkus, Environment Subcommittee Hearing, "DOE's Role in Managing Civilian Radioactive Waste," June 1, 2011, http://republicans.energycommerce.house.gov/Media/file/Hearings/ Environment/060111/Shimkus.pdf.

[54] House Energy and Commerce Subcommittee on Environment and the Economy, "Bipartisan Concern Over Administration's Haste to Terminate Permanent Nuclear Repository," press release, June 15, 2011, http://energycommerce.house.gov/news/PRArticle.aspx?NewsID=8710.

[55] *Yucca Mountain: The Administration's Impact on U.S. Nuclear Waste Management Policy*, Report by the Majority Staff of the House Science, Space, and Technology Committee, June 2011, p. 34, http://science.house.gov/press-release/committee-releases-report-outlining-administration%E2%80%99s-actions-undermine-us-nuclear.

[56] Statutory definitions for "spent nuclear fuel," "high-level radioactive waste," and "low-level radioactive waste" can be found in Section 2 of the Nuclear Waste Policy Act of 1982 (42 U.S.C. 10101). "Transuranic waste" is defined in Section 11ee. of the Atomic Energy Act (42 U.S.C. 2014); Section 11e.(2) of the act includes uranium mill tailings in the definition of "byproduct material." "Mixed waste" consists of chemically hazardous waste as defined by EPA regulations (40 CFR Part 261, Subparts C and D) that contains radioactive materials as defined by the Atomic Energy Act.

[57] Nuclear Regulatory Commission, "Staff Evaluation and Recommendations for Japan Lessons-Learned Tier 3 Issue on Expedited Transfer of Spent Fuel," op. cit., Enclosure 1, p. 77.

[58] Gutherman Technical Services, *2012 Used Fuel Data*, January 30, 2013. Adjusted for three sites closed during 2013. Includes General Electric's spent fuel storage facility at Morris, IL, located adjacent to the Dresden nuclear plant. Also, the Hope Creek and Salem nuclear plants in New Jersey are counted as a single site.

[59] Gutherman Technical Services, op. cit.

[60] DOE Office of Civilian Radioactive Waste Management, *OCRWM Annual Report to Congress, Fiscal Year 2002*, DOE/RW-0560, October 2003, Appendix C.

[61] DOE Office of Civilian Radioactive Waste Management, *Draft Supplemental Environmental Impact Statement for a Geologic Repository for the Disposal of Spent Nuclear Fuel and High-Level Radioactive Waste at Yucca Mountain, Nye County, Nevada*, Summary, DOE/EIS-0250F-S1D, October 2007, p. S-47.

[62] Gutherman Technical Services, op. cit. In addition, GE operates an independent pool storage facility near Morris, IL.

[63] Nuclear Regulatory Commission, *Waste Confidence Decision Update*, 75 *Federal Register* 81037, December 23, 2010, http://www.federalregister.gov/articles/2010/12/23/2010-31637/waste-confidence-decision-update#p-372.

[64] National Academy of Sciences, *Safety and Security of Commercial Spent Nuclear Fuel Storage: Public Report*, released April 6, 2005, p. 2.

[65] U.S. Nuclear Regulatory Commission, Near-Term Task Force Review of Insights from the Fukushima Dai-ichi Accident, *Recommendations for Enhancing Reactor Safety in the 21st Century*, p. 46, http://pbadupws.nrc.gov/docs/ ML1118/ML111861807.pdf.

[66] Nuclear Regulatory Commission, "NRC to Issue Orders, Information Request as Part of Implementing Fukushima-Related Recommendations," press release, March 9, 2012, http://www.nrc.gov/reading-rm/doc-collections/news/2012/ 12-023.pdf.

[67] Nuclear Regulatory Commission, "Staff Evaluation and Recommendations for Japan Lessons-Learned Tier 3 Issue on Expedited Transfer of Spent Fuel," *op.cit.*

[68] U.S. Department of Energy, Management Information Manifest System, http:// mims.apps.em.doe.gov/mims.asp#.

[69] Low Level Waste Forum, "WCS Announces Opening of Texas Compact Disposal Facility," November 10, 2011.

[70] BRC Final Report, op. cit.

[71] Elaine Hiruo and Herman Wang, "New Mexico Interested in Facilities for Waste Storage, Disposal: Officials," *NuclearFuel*, March 5, 2012, p. 3.

[72] Maddox, James M., Chairman, Eddy-Lea Energy Alliance, LLC, "Notice of Intent to Submit a License Application for Consolidated Used Nuclear Fuel Storage Facility," February 26, 2013, http://pbadupws.nrc.gov/docs/ML1306/ ML13067A278.pdf.

[73] Housley Carr and Elaine Hiruo, "Group Urges Mississippi to Become Home to Spent Fuel Facilities," *NuclearFuel*, September 2, 2013.

[74] DOE, *Strategy for the Management and Disposal of Used Nuclear Fuel and High-Level Waste*, op. cit.

[75] Bureau of Indian Affairs, *Record of Decision for the Construction and Operation of an Independent Spent Fuel Storage Installation (ISFSI) on the Reservation of the Skull Valley Band of Goshute Indians (Band) in Tooele County, Utah*, September 7, 2006.

[76] Bureau of Land Management, *Record of Decision Addressing Right-of-Way Applications U 76985 and U 76986 to Transport Spent Nuclear Fuel to the Reservation of the Skull Valley Band of Goshute Indians*, September 7, 2006.

[77] Winslow, Ben, "Goshutes, PFS Sue Interior," *Deseret Morning News*, July 18, 2007.

[78] U.S. District Court for the District of Utah, *Skull Valley Band of Goshute Indians and Private Fuel Storage v. United States Department of the Interior*, Civil Action No. 07-cv-0526-DME-DON, July 26, 2010, http://64.38.12.138/docs/ court/goshute/order072610.pdf.

[79] Palmberg, Robert M., Chairman of the Board, Private Fuel Storage LLC, letter to Nuclear Regulatory Commission, December 20, 2012, http://pbadupws.nrc.gov/docs/ML1235/ ML12356A063.pdf.

[80] See, for example: NRC, "Analysis of Energy Policy Act of 1992 Issues Related to High-Level Waste Disposal Standards, SECY-93-013, January 25, 1993, attachment p. 4.

[81] National Research Council. *Technical Bases for Yucca Mountain Standards*. National Academy Press. 1995.

[82] *Nuclear Energy Institute v. Environmental Protection Agency*, U.S. Court of Appeals for the District of Columbia Circuit, No. 01-1258, July 9, 2004.

[83] Especially high doses at the upper end of the exposure range would raise the mean, or average, more than the median, or the halfway point in the data set.

[84] Office of the Governor, Agency for Nuclear Projects. *Comments by the State of Nevada on EPA's Proposed New Radiation Protection Rule for the Yucca Mountain Nuclear Waste Repository*. November 2005.

[85] Posted on the EPA website at http://www.epa.gov/radiation/yucca.

[86] Nuclear Regulatory Commission, "Implementation of a Dose Standard After 10,000 Years," 74 *Federal Register* 10811, March 13, 2009.

[87] FSEIS, p. S-42. Posted on the DOE website at http://www.rw.doe.gov/ym_repository/seis/docs/002_Summary.pdf.

[88] E-mail from Thomas Santaniello, Bureau of the Fiscal Service, March 29, 2013.

[89] DOE, Office of Civilian Radioactive Waste Management, Office of Program Management, *Monthly Summary of Program Financial and Budget Information*, as of January 31, 2010, available at http://www.thenwsc.org/ym/DOE%20Financial%20&%20Budget%20Summary%20013110.pdf.
The report notes that some figures may not add due to independent rounding.

[90] Available on the DOE website at http://energy.gov/sites/prod/files/gcprod/documents/FY_2007 _TotalSystemLifeCycleCost_Pub2008.pdf.

[91] Waste Control Specialists LLC, "Historic Texas Compact Disposal Facility Ready for Business," http://www.wcstexas.com.

[92] Bionomics, Inc., "Bionomics Makes First Shipment to Texas Low Level Waste Site," press release, April 27, 2012, http://www.bionomics-inc.com/documents/Newsletter/First%20Shipment%20to%20Texas.pdf.

[93] Conley, Maureen, "Texas Commission Approves Import of LLW to New Disposal Facility," *Nucleonics Week*, March 29, 2012, p. 8.

[94] TCEQ website: http://www.tceq.state.tx.us/permitting/radmat/licensing/wcs_license_app.html#wcs_status.

[95] *Alabama et al. v. North Carolina*, S. Ct. (2010), http://www.supremecourt.gov/opinions/09pdf/132Orig.pdf.

[96] Lovell, David L., Wisconsin Legislative Council Staff, *State Statutes Limiting the Construction of Nuclear Power Plants*, October 5, 2006.

In: Federal Disposal Responsibilities
Editor: Tara C. Waller
ISBN: 978-1-63321-217-6

Chapter 4

DEFENSE SURPLUS EQUIPMENT DISPOSAL: BACKGROUND INFORMATION*

Valerie Bailey Grasso

SUMMARY

The Department of Defense (DOD) through a Defense Logistics Agency (DLA) component called DLA Disposition Services [formerly the Defense Utilization and Marketing Service (DRMS)] has a policy for disposing of government equipment and supplies considered surplus or deemed unnecessary, or excess to the agency's currently designated mission. DLA Disposition Services is responsible for property reuse (including resale), precious metal recovery, recycling, hazardous property disposal, and the demilitarization of military equipment. The effort to dispose of surplus military equipment dates back to the end of World War II when the federal government sought to reduce a massive inventory of surplus military equipment by making such equipment available to civilians. (The disposal of surplus real property, including land, buildings, commercial facilities, and equipment situated thereon, is assigned to the General Services Administration, Office of Property Disposal.)

* This is an edited, reformatted and augmented version of a Congressional Research Service publication RS20549, prepared for Members and Committees of Congress dated April 29, 2014.

BACKGROUND

On September 12, 1972, the Defense Property Disposal Service (the forerunner to DRMS) was established under the Defense Supply Agency (now DLA). Defense property disposal offices were established worldwide on or near major military installations. DLA Disposition Services is responsible for property reuse (including the disposal and sale of surplus and excess defense equipment and supplies), precious metal recovery, recycling, hazardous property disposal, and the demilitarization of military equipment. During FY2008 over $2.2 billion of property was reutilized.[1]

DLA Disposition Services provides support at major U.S. military installations worldwide. Headquartered in Battle Creek, MI, the DLA Disposition Services personnel serve in 16 foreign countries (including the Middle East and Southwestern Asia), 2 U.S. territories (Guam and Puerto Rico), and 41 states. DLA Disposition Services are provided in field offices in Afghanistan, Iraq, and Kuwait with teams deploying out to forward operating bases to assist combat units. With over 90 field offices, DLA Disposition Services employs about 1,500 people.

The Reutilization/Transfer/Donation Program establishes a process for inventory considered no longer needed by DOD to be redistributed among various groups.[2] Property disposal means redistributing, transferring, donating, selling, demilitarizing, destroying, or other "end of life cycle" activities. Disposal is the final stage before the property leaves DOD's control.[3] In some cases, the act of demilitarization—destroying the item's military offensive and defensive capability—accomplishes the intent of disposal.

Property is considered excess when one particular federal agency determines it is not needed for its particular use, while property is considered surplus when it is no longer needed by the federal government. Most property turned in to DLA Disposition Services by the military services is offered for use in other DOD activities and to other federal agencies.

Property considered surplus can be reused, transferred, donated, or sold; potential recipients may include law enforcement agencies, school systems, medical institutions, civic and community organizations, libraries, homeless assistance providers, state and local government agencies, and the public. During FY2008, about 56,000 military organizations and components turned in over 3.5 million items to DLA Disposition Services.[4] About half of all surplus items are designated for the foreign military sales program, and about half are made available to other government agencies, eligible donees, or sold to the public.[5]

On July 31, 2008, DLA awarded Liquidity Services, Inc. a contract to be the primary manager for the receipt, storage, marketing, and disposition of virtually all usable defense surplus property approved by DOD for sale to the public. The contract had a base term of three years with two one-year renewal options.[6] The contract was later extended through February 15, 2013.[7] On September 13, 2012, Liquidity Services announced that DLA had exercised the second of (potentially) two, 12-month extension periods, under its "Surplus Useable Property Sales Contract" to sell DOD surplus property. The surplus contract's performance period was extended through February 13, 2014.[8]

MAJOR NEW DEVELOPMENTS

P.L. 112-239 (H.R. 4310, 112th Congress) contains at least three provisions that impact the policy governing the distribution of DOD surplus or excess property.[9] Section 1051 expands the Secretary of the Army's authority to loan or donate small firearms, determined to be excess, for use during funerals and other ceremonial purposes;[10] Section 1053 grants the Secretary of Defense the authority to transfer mine-resistant, ambush-protected vehicles and spare parts, to nonprofit U.S. humanitarian, demining organizations for training purposes,[11] and Section 1091 grants DOD the authority to transfer certain aircraft, with exceptions, to the Secretary of Agriculture and the Secretary of Homeland Security for use by the Forest Service and the U.S. Coast Guard.[12]

CONTROLLING LEGAL AUTHORITY

Authority for the disposal of surplus defense property can be found in P.L. 94-519, 10 U.S.C. 381, which amends the Federal Property and Administrative Services Act of 1949 (40 U.S.C. 484), P.L. 107-117, and Department of Defense (DOD) Directive 4140.1, Supply Chain Materiel Management Policy; DOD 4160.21-M Defense Materiel Disposition Manual, and DOD 4160.21-M-1 Defense Demilitarization Manual.[13]

DLA Disposition Services

DLA Disposition Services manages the reutilization, transfer, donation and sale of surplus military property. The Reutilization/Transfer/Donation Program through DLA Disposition Services establishes a process for property, considered no longer needed by DOD, to be redistributed among various groups. Property considered surplus can be reused, transferred, donated, or sold; potential recipients may include law enforcement agencies, school systems, medical institutions, civic and community organizations, libraries, homeless assistance providers, state and local government agencies, veteran's organizations, and the public.

Finally, DLA Disposition Services manages the DOD surplus property sales program. Property that is no longer needed by the government may be acquired through public sales, if the property is appropriate and safe for sale to the general public.

Past Legislative Activity

P.L. 112-81 (H.R. 1540, 112th Congress) contained a provision (Section 361) that clarified a previously enacted provision (Section 346 of the Ike Skelton National Defense Authorization Act for FY2011, P.L. 111-383) which made available for sale any small arms ammunition and small ammunition components which were in excess of military requirements. Section 361 amended the conditions that would govern the commercial sale of small arms ammunition components and fired cartridge cases.

H.Rept. 112-329, the conference report that accompanied H.R. 1540, offered the following clarification.

> The conferees note that the intent of section 346 of P.L. 111-383, as amended, is to clarify that the only fired cartridge cases (referred to as expended small arms cartridge cases) subject to the provision are intact expended small arms cartridge cases and that the provision does not apply outside the continental United States or overrides established Department of Defense (DOD) trade security controls or explosives safety controls. The conferees note that the DOD would be permitted to demilitarize and recycle expended small arms cartridge cases covered by the provision so long as there is not a significant decrease in intact expended small arms cartridge cases being made available for sale and there is no evidence that

> commercial demands are not generally being met. The conferees note that based on its current force structure and training requirements, the DOD currently makes approximately 6- 8 million pounds of intact (non-demilitarized) expended small arms cartridge cases available each year for commercial sales. The conferees recognize that the amount made available may change as the DOD's force structure or training requirements change. The conferees note that the DOD would be responsible for assessing commercial demands for the purpose of implementing this requirement; the conferees understand that the DOD may choose to conduct market surveys or studies to assess commercial demands for this purpose.[14]

In the 111th Congress, the Ike Skelton National Defense Authorization Act for FY2011 (P.L. 111- 383, H.R. 6523) contained a provision (Section 1072) that amended Title 10 Section 2576 to broaden the categories of state and local organizations that would be eligible for surplus military equipment to include state and local law enforcement, firefighting, homeland security, and emergency management agencies.[15] Also, the Affordable Reloaded Munitions Supply (ARMS) Act of 2009 (H.R. 2193) was introduced in the House on April 30, 2009. The bill would have prohibited the Secretary of Defense from implementing any policy that prevents or places undue restriction on the sale of "intact spent military small arms ammunition casings" to certain domestic suppliers. The bill had 41 co-sponsors, and was referred to the House Armed Services Committee.

MAJOR RECIPIENTS

State and Local Governments

If property cannot be reused or transferred to other federal agencies, it may be donated to state and local government programs. Each state has designated a "State Agency for Surplus Property Program," a local governing authority to receive and distribute all federal surplus property. The program authorizes "screeners" to handle the logistics, and the state agency may charge a fee for handling the transaction. Eligible recipients include, but are not limited to, organizations that promote public health, safety, education, recreation, conservation, and other public needs, including veterans groups and

Native American organizations. Groups that qualify as a "service education activity" may have a slight priority in the screening process.

Law Enforcement Support Office (LESO)

LESO administers 10 U.S.C. Section 2576a, which transfers excess DOD equipment to federal and state law enforcement agencies through the 1033 Program.[16] DLA estimates that since 1990, more than $4.2 billion worth of property has been transferred; in FY2011 alone, a record $502 million worth of property was transferred.[17]

In addition, the 1122 Program (FY1994 National Defense Authorization Act) authorizes state and local governments to purchase law enforcement equipment for counter-drug activities.[18] Each state appoints a "Point of Contact (POC)" for this program. The POC may purchase items from any of the four inventory control points managed by DLA. To order items, applicants are to contact their State Agency for Surplus Property Program.

Firefighting Support Program

Title 10 U.S.C. 2576b authorizes the U.S. Department of Agriculture's Forest Service to manage the DOD firefighting property transfers. An interagency agreement between DOD and the Forest Service is under the authority of the Economy Act, 31 U.S.C. 1535. The Federal Excess Personal Property (FEPP) Program, administered through the Forest Service, provides equipment to state and territorial forestry programs for wild land and rural firefighting.[19]

Humanitarian Assistance Program

Title 10 U.S.C. 2557 authorizes DOD to provide excess property for humanitarian relief, domestic emergency assistance, and homeless veterans' assistance, as coordinated through the Defense Security Cooperation Agency, Office of Humanitarian and Refugee Affairs.[20]

Public Sales

Property not reused, transferred, or donated can be sold to the general public through public auctions and sealed bidding. Munitions, explosives, and strategic items are not sold.

Veteran Groups

There are at least two ways that veteran groups can qualify for eligibility for DOD surplus property.

If property cannot be reused or transferred to other federal agencies, it may be donated to state and local government programs. Each state has designated a "State Agency for Surplus Property Program," a local governing authority to receive and distribute all federal surplus property. The program authorizes "screeners" to handle the logistics, and the state agency may charge a fee for handling the transaction. Eligible recipients include, but are not limited to, organizations that promote public health, safety, education, recreation, conservation, and other public needs, including veteran's groups and Native American organizations. Groups that qualify as a "service education activity" may have a slight priority in the screening process.[21]

Property can also be acquired in accordance with Public Law 80-421, which authorizes the Secretaries of the Military Departments to donate or loan certain types of surplus military equipment to recognized, selected recipients.[22] The DLA Disposition Services website lists the following organizations as authorized to acquire, through donation or loan, obsolete or condemned combat material, books, manuscripts, works of art, drawings, plans and models for historical, ceremonial and display purposes:

- Veteran organizations;
- Soldiers Monument Associations;
- State Museums;
- Incorporated Museums;
- Municipal Museums, and
- Sons of Veterans Reserves.[23]

Other Programs

There are other programs which may assist organizations which may fail to qualify for DLA Disposition Services property as a DOD, federal, or donation customer. Such programs include (but are not limited to) museums, educational institutions, National Guard and Senior Reserve Officer Training Corps units, and the Civil Air Patrol.[24]

For Additional Information

The DLA Customer Contact Center is open 24 hours a day, 7 days a week, at (877) 352-2255, or at DLAContactCenter@dla.mil.For more information about DLA Disposition Services, see http://www.dispositionservices.dla. mil/ index.shtml.

End Notes

[1] From the DLA Disposition Services website at http://www.drms.dla.mil/about.shtml.

[2] https://www.dispositionservices.dla.mil/rtd03/faq.shtml.

[3] See the Acquisition Community Connection at the Defense Acquisition University, at https://acc.dau.mil.

[4] From the DLA Disposition Services website at http://www.drms.dla.mil/about.shtml.

[5] For further discussion of excess defense property, and the demilitarization and disposal of surplus military equipment, see CRS Report RL31675, Arms Sales: Congressional Review Process, by Paul K. Kerr; CRS Report RS20428, Excess Defense Articles: Grants and Sales to Allies and Friendly Countries, by Richard F. Grimmett; and CRS Report RL31686, Demilitarization of Significant Military Equipment, by Valerie Bailey Grasso. Another source for information is the Demilitarization and Disposal section of the Acquisition Community Connection at the Defense Acquisition University.

[6] http://www.liquidityservicesinc.com/company/news/press-releases/1183101/.

[7] Clabaugh, Jeff. Liquidity Services Get DOD Extension. Washington Business Journal, August 22, 2011. Accessed online at http://www.bizjournals.com/washington/news/2011/08/22/ liquidity-services-gets-dod-extension.html; also Jamil, Aisha. DOD Extends Liquidity Services Contract for Surplus Usable Property Sales. Washington Exec, August 30, 2011. Accessed online at http://www.washingtonexec.com/2011/08/dod-extends-liquidity-services-inc-contract-forsurplus-useable-property-sales/.

[8] Press Release. Liquidity Services, Inc. Receives Contract Extension for Surplus Used Property Sales, September 12, 2012. Accessed online at http://www.liquidityservicesinc.com/ company/news/press-releases/1735034/. Property is offered for sale through Liquidity Services' online auction marketplace, http://www.govliquidation.com/.

[9] H.R. 4310 became P.L. 112-239 on January 2, 2013.

[10] Section 1072. Expansion of Authority of the Secretary of the Army to loan or donate excess small arms for funeral and other ceremonial purposes. H.R. 4310, Subtitle G., Miscellaneous Authorities and Limitations.

[11] Section 1075. Authority to Transfer Surplus Mine-Resistant Ambush-Protected Vehicles and Spare Parts. H.R. 4310, Subtitle G., Miscellaneous Authorities and Limitations.

[12] H.R. 4310, Section 1091. Transfer Excess Aircraft to Other Departments of the Federal Government.

[13] http://www.acq.osd.mil/log/sci/mat_disposition.htm.

[14] Subtitle G – Other Matters. H.Rept. 112-329, H.R. 1540.

[15] Title 10 U.S.C. 2576, Surplus Military Equipment: Sale to State and Local Law Enforcement, Firefighting, Homeland Security, and Emergency Management Agencies and Firefighter Agencies.

[16] https://www.dispositionservices.dla.mil/rtd03/leso/.

[17] See LESO website at https://www.dispositionservices.dla.mil/rtd03/leso/.

[18] http://www.dla.mil/logisticsoperations/DHS/Pages/DLA1122.aspx.

[19] http://www.fs.fed.us/fire/partners/fepp/.

[20] See https://www.dispositionservices.dla.mil/rtd03/miscprograms.shtml; also, http://www.state.gov/t/pm/iso/ c21542.htm.

[21] https://www.dispositionservices.dla.mil/rtd03/index.shtml.

[22] See 10 U.S.C. 2572, Documents, Historical Artifacts, and Condemned or Obsolete Combat Materiel: Loan, Gift or Exchange. Accessed online at http://www.law.cornell.edu/uscode/text/10/2572.

[23] https://www.dispositionservices.dla.mil/rtd03/veterans.shtml.

[24] For further information, see https://www.dispositionservices.dla.mil/rtd03/index.shtml.

In: Federal Disposal Responsibilities
Editor: Tara C. Waller

ISBN: 978-1-63321-217-6

Chapter 5

BASE REALIGNMENT AND CLOSURE (BRAC): TRANSFER AND DISPOSAL OF MILITARY PROPERTY*

R. Chuck Mason

SUMMARY

The Defense Base Closure and Realignment Act of 1990 (P.L. 100-526) and the Federal Property and Administrative Services Act of 1949 (P.L. 81-152) provide the basic framework for the transfer and disposal of military installations closed during the base realignment and closure (BRAC) process. In general, property at BRAC installations is first subjected to screening for use by the Department of Defense and by other federal agencies. If no federal use for the property can be found or if an application for transfer is rejected, the property is deemed "surplus" to the needs of the federal government and made available for disposal through other mechanisms.

At this point, BRAC property is subjected to two simultaneous evaluation processes: the redevelopment planning process performed by a local redevelopment authority composed of various interested representatives of the community affected by the BRAC action; and a Department of Defense analysis prepared under the aegis of the National

* This is an edited, reformatted and augmented version of a Congressional Research Service publication R40476, prepared for Members and Committees of Congress dated February 28, 2013.

Environmental Policy Act and, eventually, informed by the local redevelopment plan.

As a part of this process, screening of the property must be performed to determine if a homeless assistance use would be appropriate. There are also a variety of "public benefit transfers," under which the property may be conveyed for various specified public purposes at reduced cost. It is also possible to dispose of BRAC property through the use of a public auction or negotiated sale, for which fair market value or a proxy for fair market value must generally be obtained. Finally the law governing the BRAC process authorizes economic development conveyances, through which a local redevelopment authority may obtain the property for specified purposes, sometimes for no consideration.

The BRAC property transfer process has been altered, both legislatively and administratively, throughout the numerous authorized closure rounds. Most recently, the National Defense Authorization Act for Fiscal Year 2010 (P.L. 111-84) amended the law with respect to economic development conveyances at no cost to local redevelopment authorities. This report provides an overview of the various authorities available under the current law and describes the planning process for the redevelopment of BRAC properties.

INTRODUCTION

The nation's military installations have gone through several rounds of base realignments and closures (BRAC), the process by which excess military facilities are identified and, as necessary, transferred to other federal agencies or disposed of, placing ownership in non-federal entities. Since the enactment of the Defense Base Closure and Realignment Act of 1990, as amended (Base Closure Act), transfer or disposal of former military installations has been governed by relatively consistent legal requirements.

On December 28, 2001, a round of base closures was authorized by Congress.[1] The BRAC process requires the Secretary of Defense to prepare and submit a list of military installations recommended for closure or realignment to the congressional defense committees and an independent commission.[2] The independent BRAC Commission, created by the Base Closure Act,[3] is required to review and analyze the Department of Defense's (DOD) recommendations and submit a report to the President with findings and conclusions that accept, reject, and/or modify the recommendations.[4] The President reviews the BRAC Commission report, and upon acceptance of the recommendations, submits it to Congress.[5] If the President fails to submit the recommendations to Congress within the timeframe required under the Base Closure Act, the BRAC process is terminated.[6] Upon receipt of the report from the President, Congress has the opportunity to disapprove of the recommendations through the enactment of a joint resolution.[7] The 2005 BRAC Commission considered 190 separate DOD recommendations, a number exceeding the number of recommendations considered by all previous BRAC Commissions combined.[8] Ultimately, the BRAC Commission recommended a total of 182 closures or realignments with an estimated savings to the taxpayer of $15 billion over 20 years.[9] The recommendations were accepted by the President and forwarded to Congress.[10] Congress did not disapprove of the report and, therefore, the recommendations became law on November 9, 2005.[11]

The current BRAC law is similar to the original statute and retains many of the transfer authorities that were available in previous rounds. Significant amendments in 1999 and 2001 altered portions of the law's disposal authorities, including requirements related to economic development conveyances. Consequently, DOD promulgated new regulations to implement the property disposal authorities available for the 2005 round.[12] However, in 2009, Congress amended the law as it relates to economic development conveyances requiring DOD to issue revised regulations. This report provides an overview of the transfer and disposal authorities available under the law for military installations closed during the 2005 round, and indicate how amendments to the Base Closure Act have altered the property transfer and disposal process.[13] It also describes DOD's regulations implementing the amended Base Closure Act.

TRANSFER, DISPOSAL, AND LEASING AUTHORITIES

The transfer or disposal of federal property is primarily performed by the General Services Administration (GSA) pursuant to the Federal Property and Administrative Services Act of 1949 (FPASA).[14] The Base Closure Act directs the Administrator of the GSA to delegate specified transfer and disposal authorities to DOD for use at BRAC installations, and DOD has, in turn, delegated this authority to the various military departments.[15] Thus, BRAC property transfer and disposal is performed, generally, in accordance with the FPASA and the GSA regulations implementing it. In addition, the Base

Closure Act authorizes DOD, with GSA approval, to supersede GSA regulations with BRAC-specific regulations.[16]

Apart from the transfer and disposal authorities typically available for federal property, the Base Closure Act and other provisions of law authorize a variety of other conveyance mechanisms. The available authorities include: public benefit transfers, economic development conveyances (at cost and no cost), negotiated sales to state or local governments, conservation conveyances, and public sales.[17] In some cases, the analysis and use of particular authorities must precede analysis and use of others. On the other hand, there are many transfer and disposal mechanisms that are given roughly equivalent priority; thus analysis and use of them may occur simultaneously.

In addition to DOD's role in making disposal and transfer determinations, the Base Closure Act also provides a substantial role for states and communities in the property redevelopment planning process. Thus, local communities can significantly affect the BRAC property transfer and disposal decisions made at the federal level. The specific roles for states and communities as well as the various transfer and disposal authorities are discussed below.

Local Redevelopment Authorities (LRAs)

Pursuant to the act, an LRA is "any entity (including an entity established by a State or local government) recognized by the Secretary of Defense as the entity responsible for developing the redevelopment plan with respect to the installation or for directing the implementation of such plan."[18] DOD must prepare an environmental impact analysis under the National Environmental Policy Act (NEPA), in which it must examine all reasonable disposal alternatives and make its own disposal decisions.[19] However, LRAs are responsible for designing a comprehensive plan for reuse of BRAC property, culminating in a redevelopment plan, which is submitted to DOD and included as part of the proposed federal action.[20] While the redevelopment plan is not binding on DOD, it may have significant influence on its disposal decisions, and, in some instances, DOD is statutorily directed to give the plan considerable weight.[21] Local zoning authorities and state land use regulations may also impact the disposal decisions made by DOD.

The Base Closure Act does not establish statutory requirements for the formation of LRAs. DOD regulations provide that the LRA should have "broad-based membership, including, but not limited to, representatives from those jurisdictions with zoning authority over the property."[22] The regulations further state that "[g]enerally, there will be one recognized LRA per installation."[23] In the event that a LRA is not recognized by DOD, or if the LRA fails to timely submit a redevelopment plan, the Secretary concerned is required to consult with the state's Governor and heads of local governments before proceeding with the disposal of the property according to applicable laws.[24]

Transfers for Federal Utilization

It is DOD policy to act expeditiously under the BRAC process, whether it is the closing or realigning of an installation, in order to facilitate the transfer of real property for community reuse.[25] Prior to consideration of transfer to a non-federal entity, the property must be screened for continued federal use.

DOD Components or Other Agencies

The first step in the property transfer process begins when the military service in possession of a BRAC property notifies other DOD components and federal agencies that property is in "excess"26 to its needs and has become available.27 If a DOD component or other federal agency wishes to acquire BRAC property, it must "provide a written, firm expression of interest ... [and] explain the intended use and the corresponding requirement for the buildings and property" within thirty days of the notice of availability,28 followed by an application for transfer of the property.29 The application must support a variety of transfer requirements, including that the property requested be better suited to the requestor's needs than its existing property or other properties and that the transfer would not create a new government program.30 During the federal screening, the Secretary concerned is required to keep the LRA informed of the progress and to provide contact information for federal agencies so that the LRA may be involved.31 DOD components and other agencies are encouraged to include the LRA, if it exists, in discussions related to the proposed use of the property.32 Ultimately, it is the responsibility of the transferring DOD component to review the applications and make a determination as to whether the transfer is appropriate based on several factors:

- the requirement for additional property must be valid and appropriate;
- the proposed use is consistent with the highest and best use of the property;
- the proposed transfer will not have an adverse impact on the transfer of any remaining portion of the installation;
- the proposed transfer will not establish a new program or substantially increase the level of a component's or agency's existing programs;
- the application offers fair market value for the property, unless waived;
- the proposed transfer addresses applicable environmental responsibilities to the satisfaction of the Secretary concerned; and
- the proposed transfer is in the best interest of the Government.[33]

In the event multiple acceptable applications for the same piece of BRAC property are submitted, the Secretary must consider, in order:

- the need to perform the national defense missions of the Department of Defense and the Coast Guard;
- the need to support the homeland defense mission; and
- the LRA's comments as well as other factors in the determination of highest and best use.[34]

If, after consideration of the applications, a determination is made that a federal-to-federal transfer is appropriate, the transfer may occur with or without compensation.[35] However, DOD regulations require that if the property is being transferred out of DOD, "fair market value reimbursement to the Military Department" be made unless the obligation is "waived by the Office of Management and Budget and the Secretary concerned, or a public law specifically provides for a non-reimbursable transfer."[36] If the federal agency receiving the property fails to provide fair market value reimbursement, the property is to be declared "surplus"[37] and disposed of in accordance with applicable laws.[38] If no DOD components or other federal agencies pursue acquisition, or if DOD denies an application for transfer, the property is determined to be surplus and the disposal process begins.

Public Domain Lands[39]

Simultaneous to the DOD component or other agency review process, and prior to a final determination that the BRAC property is surplus, DOD must determine whether the installation includes "public domain lands."[40] If the lands comprising the closed or realigned installation were originally withdrawn from the public domain for use as a military facility, then, in accordance with FPASA, the Department of the Interior (DOI), acting through the Bureau of Land Management (BLM), may review the property and decide whether the land is suitable for return to the public domain.[41] If DOD decides it will not retain the property for one of its components, it issues a Notice of Intent to Relinquish.[42] It is then the responsibility of the BLM to determine if the land is suitable to be returned to the DOI or if it should be disposed of under the Base Closure Act.[43] Because BRAC property withdrawn from the public domain would not be listed in the notice of availability sent to DOD components and other federal agencies, is not clear whether a period for federal-to-federal transfers, as described above, would be available if BLM rejects the property.

Transfers for Non-Federal Utilization

Homeless Assistance

The Stewart B. McKinney Homeless Assistance Act[44] which allows "excess," "surplus," "unutilized," and "underutilized" federal property to be used as homeless shelters, previously applied to BRAC closures.[45] However, the Base Closure Community Development and Homeless Assistance Act of 1994 changed the process for BRAC properties closed after October 25, 1994.[46] The Secretary of Defense is required to publish notice of available property and to submit information on the property to the U.S. Department of Housing and Urban Development (HUD), as well as to the LRA for that particular installation.[47] All interested parties, including representatives of the homeless, are then to submit to the LRA a notice of interest in the property.[48] The LRA is to consider "the interests in the use to assist the homeless of the buildings and property at the installation that are expressed in the notices submitted to the redevelopment authority ..." in preparing its redevelopment plan.[49] Upon completion of its plan, the LRA submits it to the Secretary of HUD and the Secretary of Defense for review.

The Secretary of HUD is authorized to review the plan, negotiate with the LRA for changes, and based on statutorily prescribed factors determine whether the plan is acceptable.[50] Upon HUD approval, the base redevelopment plan, including any homeless assistance component and agreement to implement no cost homeless assistance property conveyances, is submitted to

DOD. DOD is required to give the redevelopment plan's homeless assistance recommendations "substantial deference."[51] The Base Closure Community Development and Homeless Assistance Act of 1994, as originally enacted, required the Secretary of Defense to dispose of the property according to the LRA plan, including any homeless assistance designations.[52] The substantial deference requirement, added by the Base Closure Act, appears to clarify DOD's authority to dispose of property in a manner inconsistent with the LRA redevelopment plan, as long as the required level of deference was afforded.[53]

Public Benefit Transfers

Public benefit transfers are authorized under FPASA and allow for conveyance of property at a discount or for no cost for specified public purposes[54] Only certain entities may acquire property through a public benefit transfer, and the categories of acceptable recipients vary according to the type of public benefit use contemplated. For instance, transfers for use in the protection of public health may be to a state, a public subdivision or instrumentality of a state, a tax-supported medical institution, or a 501(c)(3) nonprofit hospital or similar institution.[55]

DOD is required to inform the various agencies exercising authority over public benefit transfer programs of potentially available property and to inform the relevant LRA of any interest expressed by agencies.[56] The LRAs are encouraged to coordinate with interested parties and make a reasonable effort to incorporate their interests within the redevelopment plan.[57] However, there is no requirement that their interests be included in the redevelopment plan, they must only be considered by the LRA.[58] DOD is also required, through the military departments, to conduct an official public benefit transfer screening in accordance with the Federal Property Management Regulations based on potential uses indentified in the redevelopment plan. If a public transfer is made, the transferring instrument will generally contain various binding "terms, conditions, reservations, and restrictions" to ensure the use of the property for the purposes for which it was transferred.[59] The LRA is responsible for the implementation of and compliance with the legally binding terms[60] In the event the agreement is violated and the property reverts to the LRA, the LRA is responsible for ensuring the future utilization of the property.[61]

Conservation Conveyances

If BRAC property remains available after it has been considered for both a federal-to-federal transfer and a public benefit conveyance, DOD is authorized to transfer BRAC property via a conservation conveyance.[62] To be eligible for a conservation conveyance the property must be suitable and desirable for conservation purposes, must have been made available for a public benefit transfer "for a sufficient period of time," and must not be subject to a pending request for a public benefit transfer or for transfer to another federal agency.[63] In general, a conservation conveyance is to be for reduced cost.[64] The conveyance may be made to a state or qualified nonprofit entity for conservation purposes and must be subject to a reversionary clause authorizing the United States to reclaim the property should the use for conservation purposes cease.[65] With the concurrence of the Secretary of the Interior, DOD may grant the release from a covenant restricting future conveyances, but only if fair market value for the property is paid.[66]

Public Auction and Negotiated Sale

In accordance with FPASA, DOD may dispose of BRAC property via public auction or through a negotiated sale with a single purchaser.[67] The public auction process requires public advertising for bids under terms and conditions that permit "full and free competition consistent with the value and nature of the property involved."[68] If adequate bids are received and disposal is in the public interest, the bid most advantageous to the federal government is to be accepted.[69] A negotiated sale is permissible when: (1) it is necessary in the public interest; (2) the public health, safety, or national security will be promoted by particular disposal of personal property; (3) a public exigency makes an auction unacceptable; (4) a public auction would adversely impact the national economy; (5) fair market value does not exceed $15,000; (6) a public auction has failed to produce acceptable bids; (7) the character of the property makes public auction impractical; (8) disposal is to a state, territory, or U.S. possession; or (9) negotiated sale is authorized by other law.[70]

Economic Development Conveyances (EDCs)

In addition to FPASA authorities, the Base Closure Act has since its enactment provided for EDCs in one form or another. Under its EDC authority, DOD may convey BRAC property to a LRA for less than fair market value. From 1994 until the 1999 and 2001 amendments to the Base Closure Act, the Secretary of Defense was authorized to "transfer real property and personal property located at a military installation to be closed ... to the redevelopment authority ... for consideration at or below the fair market value of the property transferred or without consideration."[71] The reduced or no cost conveyance was authorized when it was determined to be necessary to support economic development and when DOD could show that other transfer authorities were insufficient.[72]

Amendments to the Base Closure Act in 1999 and 2001 significantly altered the requirements applicable to the use of an EDC.[73] Under section 2905(b), the broad discretion of the Secretary of Defense to authorize reduced or no consideration economic development conveyances was replaced by arguably a more restrictive scheme. Among the changes, for installations closed after January 1, 2005, the Secretary was required to "seek to obtain consideration in connection with any transfer ... in an amount equal to the fair market value of the property, as determined by the Secretary."[74] However, transfers of property without consideration, in limited circumstances, were authorized. The law provided that: "the transfer of property of a military installation ... may be without consideration" only when the transferee agrees to specified terms.[75] These terms include a requirement that the recipient LRA use the proceeds from certain future sales or leases of the acquired property to support economic redevelopment at the former installation and accept control of the property "within a reasonable time after the date of the property disposal record of decision."[76]

However, in 2009, noting that many "negotiations between the Department of Defense and local redevelopment authorities ... over the value of property to be disposed under an economic development conveyance (EDC) have stalled over the past 2 years due to difficulties in the nation's financial markets, the deterioration of local economic conditions, and the potential of legislative changes,"[77] Congress further amended the Base Closure Act with respect to utilization of a no cost EDC. The requirement that the Secretary seek consideration in an amount equal to the fair market value of the property has been removed. The law now allows that the transfer of property "may be for consideration at or below the estimated fair market value or without consideration."[78] The determination of what consideration is to be received, if any, "may account for the economic conditions of the affected community and the estimated costs to redevelop the property."[79] The Secretary is authorized to accept as consideration: a share of the revenues the LRA receives from the property; goods and services; real property and improvements; or other consideration the Secretary considers appropriate.[80] The amendment does not change the requirement that the LRA use proceeds from the acquired property to support economic redevelopment.[81]

The LRA may apply for an EDC after completion of its redevelopment plan. An application must be submitted consistent with a schedule devised by the Secretary of the transferring DOD component.[82] The Secretary concerned, when practicable, provides a preliminary determination within 30 days of receipt as to whether the Military Department can accept the application for negotiation of terms and conditions.[83] The LRA application shall include a "description of how the EDC will contribute to short- and long-term job generation on the installation" and provide a "description of the economic impact of closure or realignment on the local community."[84] Further, the application shall contain a statement "describing why an EDC will more effectively enable achievement of the job generation objectives of the redevelopment plan regarding the parcel requested for conveyance than other federal real property disposal authorities."[85] The transferring Secretary is required to evaluate the application and its proposed terms and conditions in accordance with a series of prescribed factors, including the economic effects on the community of the proposed EDC, the interests and concerns of other federal agencies, and the economic benefit to the United States.[86] The regulations addressing an EDC without consideration if the LRA agrees that "proceeds from any sale or lease of the property ... during at least the first seven years ... [following transfer] shall be used to support economic redevelopment.... " do not appear to be in conflict with the amended law.[87] The authorized uses to support economic redevelopment, unchanged by the 2009 amendment, are:

- road construction;
- transportation management facilities;
- storm and sanitary sewer construction;
- police and fire protection facilities and other public facilities;
- utility construction;
- building rehabilitation;
- historic property preservation;
- pollution prevention equipment or facilities;
- demolition;
- disposal of hazardous materials generated by demolition;
- landscaping, grading, and other site or public improvements; and
- planning for or the marketing of the development and reuse of the installation.[88]

If the LRA does not utilize the funds in support of economic redevelopment, DOD is authorized under the Base Closure Act to recoup the portion of the proceeds received by the LRA in an amount it deems appropriate.[89]

Leases

In addition to the final conveyance of property contemplated by the Base Closure Act, federal law authorizes the leasing of BRAC property to both federal and non-federal lessees.

Leaseback

The law and regulations authorize what has been referred to as a "leaseback," an arrangement wherein the transferring Secretary conveys property to a LRA and the LRA agrees to lease the property to a federal agency.[90] Under the regulations, this arrangement will only be used if the agency that would lease the property agrees to the arrangement, the LRA and the agency can agree to lease terms, and the transferring Secretary determines the arrangement is in the interest of the DOD component or agency.[91] The leases are to be for terms of no more than fifty years, subject to renewal, and cannot require rental payments.[92]

Non-Federal Lessee

While the Base Closure Act does not specifically provide for the authority to lease property to non-federal lessees, it does indicate that proceeds from leases are to be deposited into a BRACspecific account. [93] The authority for non-federal leases is contained in 10 U.S.C. § 2667, the same statute governing the leasing of non-BRAC military property.[94] DOD's regulations identify that the leasing of BRAC properties prior to final disposition "may facilitate state and local economic adjustment efforts and encourage economic development, but the Secretary concerned will always concentrate on the final disposition of real and personal property."[95] Lessees must generally pay fair market value; however, less than fair market value consideration is authorized if the Secretary finds that:

- a public interest will be served as a result of the lease; and
- the fair market value of the lease is unobtainable or not compatible with such public benefit.[96]

Prior to a BRAC property being leased, the law requires DOD to consult with the Administrator of the Environmental Protection Agency (EPA) to determine whether the property is in suitable condition for leasing.[97] In general, NEPA requires federal agencies to analyze the environmental impacts of a proposed federal action and alternatives to that action.[98] The statute governing BRAC property leases indicates that the scope of environmental analysis required is "limited to the environmental consequences of activities authorized under the proposed lease and the cumulative impacts of other past, present, and reasonably foreseeable future actions during the period of the proposed lease."[99] However, this relief from full application of NEPA does not apply if activities authorized under the lease would:

- significantly affect the quality of the human environment; or
- irreversibly alter the environment in a way that would preclude any reasonable disposal alternative of the property concerned.[100]

Additional regulatory and statutory provisions indicate that leases of BRAC property are intended to be short-term, interim measures to spur economic development pending final disposition, and therefore these leases "make no commitment for future use of ultimate disposal."[101] More specifically, the regulations indicate that lease terms may extend up to five years, including renewal options, if the lease is entered into prior to completion of the final disposal decision.[102] After completion of the final disposal decisions, the lease term may be longer than five years.[103]

When a lease is to a LRA and is provided at below fair market value and the property is later subleased, the LRA is required to apply the proceeds to the "protection, maintenance, repair, improvement, and costs related to the [leased] property.... "[104]

CONCLUSION

The Base Closure Act and the FPASA primarily governed the transfer and disposal process for 2005 round BRAC properties. The process first requires screening to determine if other DOD components or federal agencies have a need for the property. In the event that property is not transferred in this manner, it is deemed surplus and may be disposed of pursuant to BRAC and FPASA authorities. Compliance with these authorities generally requires an analysis of suitability for homeless assistance or a public benefit transfer. DOD is directed to take into consideration multiple factors in determining which authority to use, including consultation with LRAs and their redevelopment plans, but DOD appears to be ultimately responsible for making final determinations. Public auctions and negotiated sales are generally available, although it would appear that fair market value must generally be obtained under these authorities. EDCs are authorized as well, which may be made for no consideration, contingent upon certain conditions of transfer.

End Notes

[1] National Defense Authorization Act for Fiscal Year 2002, Act of December 28, 2001, P.L. 107-107, 115 Stat 1012 (current version at 10 U.S.C. § 2687 note).

[2] Base Closure Act at § 2903(c).

[3] Id. at § 2902.

[4] Id. at § 2903(c).

[5] Id. at § 2903(e) (The President may disapprove all or part of the recommendations, in which case, the report is returned to the Commission. The Commission must then submit a revised list to the President. If the President approves the revised list, it is forwarded to Congress, but if the President does not approve the revised list, the BRAC process is terminated.).

[6] Id.

[7] Id. at § 2904(b).

[8] The Defense Base Closure and Realignment Commission, 2005 Defense Base Closure and Realignment Commission Report, Executive Summary, September 2005 (Available online at http://www.brac.gov).

[9] Id.

[10] In Dalton v. Specter, 511 U.S. 462 (1994), the U.S. Supreme Court held that actions by the Secretary of Defense and the BRAC Commission are not reviewable final agency actions within the meaning of the Administrative Procedure Act (APA), since their reports recommending base closings carry no direct consequences. However, the action of the President will directly affect bases and, as such, is the final action affecting the military installations; but because the President is not an agency under the APA, that action is not reviewable under the act. The Court further held that where a statute commits decision-making to the President's discretion, judicial review of his decision is not available.

[11] Donna Miles, "BRAC Deadline Expires; DOD to Begin Closures, Realignments," American Forces Press Service, News Articles, November 9, 2005.

[12] 32 C.F.R. pt. 174.

[13] It should be noted that significant issues related to environmental cleanup under the Comprehensive Environmental Response Compensation and Liability Act (CERCLA) exist at some BRAC properties and that the use of certain property transfer authorities may be contingent upon adequate performance of CERCLA obligations or agreement by the acquiring entity to accept liability for environmental cleanup. See 42 U.S.C. § 9620(h); P.L. 107-107, § 3006. For background on environmental cleanup issues and BRAC, see CRS Report RS22065, Military Base Closures: Cleanup of Contaminated Properties for Civilian Reuse, by David M. Bearden.

[14] Act of June 30, 1949, ch. 288, 63 Stat 377. Transfer and disposal authority is codified at 40 U.S.C. §§ 521-559.

[15] Base Closure Act, § 2905(b); 32 C.F.R. § 174.5.

[16] Base Closure Act, § 2905(b).

[17] 32 C.F.R. § 174.4(b).
[18] Base Closure Act, § 2910(9).
[19] 42 U.S.C. §§ 4321 et seq.
[20] 32 C.F.R. § 174.6.
[21] The specific requirements impacting the LRA planning process and DOD's eventual disposal of property are discussed in the sections of this report addressing each disposal mechanism.
[22] 32 C.F.R. § 174.6(a).
[23] Id.
[24] Base Closure Act, § 2905(b)(3)(B); 32 C.F.R. § 174.6(c)(2).
[25] 32 C.F.R. § 174.4.
[26] "Excess" property is defined as "property under the control of a federal agency that the head of the agency determines is not required to meet the agency's needs or responsibilities." 40 U.S.C. § 102(3); 32 C.F.R. § 174.3(e).
[27] 32 C.F.R. § 174.7(a), (c).
[28] Id. at § 174.7(d).
[29] Id. at § 174.7(e).
[30] Id. at § 174.7(h).
[31] Id. at § 174.7(f).
[32] Id. at § 174.7(g).
[33] Id. at § 174.7(i).
[34] 32 C.F.R. § 174.7(j).
[35] Base Closure Act; § 2905(b)(2)(C).
[36] 32 C.F.R. § 174.7(h)(8).
[37] "Surplus" property is defined as "excess property that the Administrator determines is not required to meet the needs or responsibilities of all federal agencies." 40 U.S.C. § 102(10); 32 C.F.R. § 174.3(l).
[38] 32 C.F.R. § 174.7(k).
[39] Public domain lands are lands owned by the United States for the benefit of the citizens of the United States.
[40] 32 C.F.R. § 174.7(l).
[41] Id.
[42] 32 C.F.R. § 174.7(l)(4).
[43] 32 C.F.R. § 174.7(l)(5), (6).
[44] 42 U.S.C. § 11411.
[45] Id. at § 11411(a).
[46] P.L. 103-421, 108 Stat. 4346 (1994).
[47] Base Closure Act, § 2905(b)(7); 32 C.F.R. § 176.20.
[48] 32 C.F.R. § 176.20(c).
[49] Base Closure Act, § 2905(b)(7)(F)(i).
[50] Base Closure Act, § 2905(b)(7)(H).
[51] Base Closure Act, § 2905(b)(7)(K)(iii) (while the term "substantial deference" is not further defined by the Base Closure Act or DOD regulations, judicial application of the term may be instructive. See, e.g., Chevron v. NRDC, 467 U.S. 837 (1984); Auer v. Robbins, 519 U.S. 452 (1997)).
[52] P.L. 103-421, 108 Stat. 4346 (1994).
[53] Base Closure Act, § 2905(b)(7)(K)(iii).

[54] See 40 U.S.C. §§ 541 et seq., 49 U.S.C. §§ 47151-47153 (authorized transfers include uses for airports, historic monuments, education, national service activities, public parks and recreation, low income assistance housing, and public health purposes).

[55] Id. at § 550(d).

[56] 32 C.F.R. § 176.45.

[57] 32 C.F.R. § 176.20.

[58] Id.

[59] 40 U.S.C. § 550(b).

[60] 32 C.F.R. § 176.45(d).

[61] 32 C.F.R. § 176.45(e) (a preference exists for the LRA to utilize the property to assist the homeless, but it is not a requirement).

[62] 10 U.S.C. § 2694a.

[63] Id. at § 2694a(a).

[64] Id. at § 2694a(g).

[65] Id. at § 2694a(b), (c).

[66] Id. at § 2694a(d) (under certain circumstances the Secretary may accept less than fair market value for the property).

[67] 40 U.S.C. § 545.

[68] Id. at § 545(a)(2).

[69] Id. at § 545(a)(4).

[70] Id. at § 545(b).

[71] P.L. 103-160, § 2903 (1994).

[72] Id. (Additionally, a no consideration transfer was formerly required when a closure was to take place in a rural area and would cause "a substantial adverse impact (as determined by the Secretary) on the economy of the communities in the vicinity of the installation and on the prospect for economic recovery.... " P.L. 103-160, § 2903, amended by P.L. 106-65) For a discussion of the policy behind EDC, see Randall S. Beach, Swords to Plowshares: Recycling Cold War Installations, 15 PROB. & PROP. (2001).

[73] Act of October 5, 1999, P.L. 106-65, 113 Stat. 512; P.L. 107-107, § 3006. Bases closed under previous BRAC law but still owned by the Department of Defense may be included under the new statutory framework, and certain existing contracts may be modified to comply with the updated law.

[74] Base Closure Act, § 2905(b)(4)(B); see also 32 C.F.R. § 174.9(b).

[75] P.L. 106-65, amended by P.L. 107-107.

[76] Base Closure Act, § 2905(b)(4)(B)(ii).

[77] U.S. Congress, National Defense Authorization Act for Fiscal Year 2010, Conference Report to accompany H.R. 2647, 111th Cong., 1st sess., October 7, 2009, H.Rept. 111-288 (Washington: GPO, 2009), p. 883.

[78] P.L. 111-84, § 2715 (2009).

[79] Id.

[80] Id.

[81] Id.

[82] 32 C.F.R. § 174.9(c).

[83] Id. at § 174.9(e).

[84] Id. at § 174.9(d).

[85] Id. at § 174.9(d)(5) (All elements to be addressed by the LRA in its application for an EDC are contained in 32 C.F.R. § 174.9(d)(1) – (9)).

[86] Id. at § 174.9(e). (Prior to 2009, the Secretary was required to appraise the property, utilizing the most recent edition of the Uniform Appraisal Standards for Federal Land Acquisitions, and determine its fair market value prior to accepting an application. Additionally, prior regulations, since reversed, required the Secretary to seek to obtain consideration at least equal to the fair market value as part of an EDC.).
[87] Base Closure Act, § 2905(b)(4)(i); 32 C.F.R. § 174.9(d)(8).
[88] Base Closure Act, §2905(b)(4)(C);32 C.F.R. §174.9(d)(8)(i)-(xii).
[89] Base Closure Act, §2905(b)(4)(D); 32 C.F.R. § 174.9(j).
[90] Base Closure Act, § 2905(b)(4)(E); 32 C.F.R. § 174.12.
[91] 32 C.F.R. § 174.12(f).
[92] Id. at §174.12(h).
[93] Base Closure Act, §§ 2906, 2906A; see also 10 U.S.C. § 2667(e)(5).
[94] 10 U.S.C. § 2667(g).
[95] 32 C. F.R. § 174.11(a).
[96] Id. at § 174.11(b).
[97] 10 U.S.C. § 2667(g)(3).
[98] 42 U.S.C. §§ 4321 et seq.
[99] 10 U.S.C. § 2667(g)(4)(A).
[100] Id. at § 2667(g)(4)(C).
[101] 32 C.F.R. § 174.11(c); see also 10 U.S.C. § 2667(g)(4)(B).
[102] 32 C.F.R. § 174.11(c).
[103] Id.
[104] Id. at § 174.11(d).

INDEX

A

B

C

D

G

H

I

J

O

P

Q

R

S

Y

Z